SpringerBriefs in Statistics

SpringerBriefs present concise summaries of cutting-edge research and practical applications across a wide spectrum of fields. Featuring compact volumes of 50 to 125 pages, the series covers a range of content from professional to academic. Typical topics might include:

- A timely report of state-of-the art analytical techniques
- A bridge between new research results, as published in journal articles, and a contextual literature review
- A snapshot of a hot or emerging topic
- An in-depth case study or clinical example
- A presentation of core concepts that students must understand in order to make independent contributions

SpringerBriefs in Statistics showcase emerging theory, empirical research, and practical application in Statistics from a global author community.

SpringerBriefs are characterized by fast, global electronic dissemination, standard publishing contracts, standardized manuscript preparation and formatting guidelines, and expedited production schedules.

More information about this series at https://link.springer.com/bookseries/8921

Louis J. M. Aslett · Frank P. A. Coolen ·
Jasper De Bock

Editors

Uncertainty in Engineering

Introduction to Methods and Applications

 Springer

Editors
Louis J. M. Aslett
Department of Mathematical Sciences
Durham University
Durham, UK

Frank P. A. Coolen
Department of Mathematical Sciences
Durham University
Durham, UK

Jasper De Bock
Foundations Lab for imprecise probabilities
Ghent University
Zwijnaarde, Belgium

ISSN 2191-544X   ISSN 2191-5458 (electronic)
SpringerBriefs in Statistics
ISBN 978-3-030-83639-9   ISBN 978-3-030-83640-5 (eBook)
https://doi.org/10.1007/978-3-030-83640-5

This Springer imprint is published by the registered company Springer Nature Switzerland AG
The registered company address is: Gewerbestrasse 11, 6330 Cham, Switzerland

Preface

This book results from the Second Training School of the European Research and Training Network UTOPIAE: Uncertainty Treatment and Optimisation in Aerospace Engineering (www.utopiae.eu), at Durham University (United Kingdom) from 2 to 6 July 2018. The main focus was on uncertainty quantification.

This book consists of 9 chapters providing introductory overviews to topics on uncertainty quantification in engineering, written by topic experts who presented lectures at the school. The book starts with introductions to Bayesian statistics and Monte Carlo methods, followed by three chapters on imprecise probabilities: introductions to the general theory and to imprecise Markov chains, and a short overview of statistical methods. Then attention shifts to reliability theory and simulation methods for complex systems. The final two chapters are on aspects of aerospace engineering, considering stochastic model updating from an imprecise Bayesian perspective, and uncertainty quantification for aerospace flight modelling.

We are grateful to the chapter authors for their contributions to this book, and for their enthusiastic presentations at the training school. We also thank Massimiliano Vasile for leading the UTOPIAE network and bringing experts from different fields together to enhance research in the fascinating field of aerospace engineering. We thank the staff at Springer for their support in the preparation of this book for the Springer Briefs in Statistics series.

This book is sponsored by the UTOPIAE ITN, and the corresponding Training School was also organised within the UTOPIAE ITN, supported by H2020-MSCA-ITN-2016 UTOPIAE, grant agreement 722734.

Durham, United Kingdom   Louis J. M. Aslett
Ghent, Belgium   Frank P. A. Coolen
October 2020   Jasper De Bock

Contents

Chapter 1
Introduction to Bayesian Statistical Inference

Georgios P. Karagiannis

Abstract We present basic concepts of Bayesian statistical inference. We briefly introduce the Bayesian paradigm. We present the conjugate priors; a computational convenient way to quantify prior information for tractable Bayesian statistical analysis. We present tools for parametric and predictive inference, and particularly the design of point estimators, credible sets, and hypothesis tests. These concepts are presented in running examples. Supplementary material is available from GitHub.

1.1 Introduction

Statistics mainly aim at addressing two major things. First, we wish to learn or draw conclusions about an unknown quantity, $\theta \in \Theta$ called 'the parameter', which cannot be directly measured or observed, by measuring or observing a sequence of other quantities called 'observations (or data, or samples)' $x_{1:n} := (x_1, \dots, x_n) \in \mathcal{X}^m$ whose generating mechanism is (or can be considered as) stochastically dependent on the quantity of interest θ though a probabilistic model $x_{1:n} \sim f(\cdot|\theta)$. This is an inverse problem since we wish to study the cause θ by knowing its effect $x_{1:n}$. We will refer to this as parametric inference. Second, we wish to learn the possible values of a future sequence of observations $y_{1:m} \in \mathcal{X}^m$ given $x_{1:n}$. This is a forward problem, and we will call it predictive inference. Here, we present how both inferences can be addressed in the Bayesian paradigm.[1]

Consider a sequence of observables $x_{1:n} := (x_1, \dots, x_n)$ generated from a sampling distribution $f(\cdot|\theta)$ labeled by the unknown parameter $\theta \in \Theta$. The statistical model $\mathfrak{m}$ consists of the observations $x_{1:n}$, and their sampling distribution $f(\cdot|\theta)$; $\mathfrak{m} = (f(\cdot|\theta); \ \theta \in \Theta)$.

---

[1] https://github.com/georgios-stats/UTOPIAE-Bayes.

G. P. Karagiannis (✉)
Department of Mathematical Sciences, Durham University, Durham, United Kingdom
e-mail: georgios.karagiannis@durham.ac.uk

© The Author(s) 2022
L. Aslett et al. (eds.), *Uncertainty in Engineering*,
SpringerBriefs in Statistics,
https://doi.org/10.1007/978-3-030-83640-5_1

Unlike in Frequentist statistics, in Bayesian statistics unknown/uncertain parameters are treated as random quantities and hence follow probability distributions. This is justified by adopting the subjective interpretation of probability [4], as the degree of the researcher's believe about the uncertain parameter θ. Central to the Bayesian paradigm is the specification of the so-called prior distributions $d\pi(\theta)$ on the uncertain parameters θ representing the degree of believe (or state of uncertainty) of the researcher about the parameter. Different researchers may specify different prior probabilities, as this is in accordance to the subjective nature of the probability. The specification of the prior is discussed in Sect. 1.2.

The Bayesian model consists of the statistical model $f(x_{1:n}|\theta)$ containing the information about θ available from the observed data $x_{1:n}$, and the prior distribution $\pi(\theta)$ reflecting the researcher's believe about θ before the data collection. It is denoted as

$$(f(x_{1:n}|\theta), \pi(\theta)) \text{ or as } \begin{cases} x_{1:n}|\theta & \sim f(\cdot|\theta) \\ \theta & \sim \pi(\cdot) \end{cases}.$$

Bayesian parametric inference relies on the posterior distribution $\pi(\theta|x_{1:n})$ whose density or mass function (PDF or PMF) is calculated by using the Bayes theorem

$$\pi(\theta|x_{1:n}) = \frac{f(x_{1:n}|\theta)\pi(\theta)}{\int_\Theta f(x_{1:n}|\theta)\pi(d\theta)} \tag{1.1}$$

as a tool to invert the conditioning from $x_{1:n}|\theta$ to $\theta|x_{1:n}$. Posterior distribution (1.1) quantifies the researcher's degree of believe after taking into account the observations. By using subjective probability arguments, we can see interpret (1.1) as a mechanism that updates the researcher's degree of believe from the prior $\pi(\theta)$ to the posterior $\pi(\theta|x_{1:n})$ in the light of the observations collected.

Bayesian predictive inference about a future observation y_* can be addressed based on the predictive distribution defined as

$$p(y|x_{1:n}) = \int_\Theta f(y|\theta)\pi(d\theta|x_{1:n}) = E_\pi(f(y|\theta)|x_{1:n}). \tag{1.2}$$

Essentially, it is the expected value of the sampling distribution averaging out the uncertain parameter θ with respect to its posterior distribution reflecting the researcher's degree of believe.

Although the posterior and predictive distributions quantify the researcher's knowledge, they are not enough to give a solid answer about the quantity to be learned. In what follows we discuss important concepts based on decision theory which are used for Bayesian inference.

1.2 Specification of the Prior

Prior distribution $\pi(\theta)$ needs to reflect the researcher's degree of believe about the uncertain parameter $\theta \in \Theta$. Sophisticated prior distributions often lead to ineluctable posterior or predictive probabilities, and hence Bayesian analysis. Following, we present a computationally convenient class of priors applicable to several scenarios.

1.2.1 Conjugate Priors

Conjugate priors is a mathematically convenient way to specify the prior model in certain cases. They facilitate the tractable implementation of the Bayesian statistical analysis, by leading to computationally tractable posterior distributions.

Formally, if $\mathcal{F} = \{f(\cdot|\theta); \forall \theta \in \Theta\}$ is a class of parametric models (sampling distributions), and $\mathcal{P} = \{\pi(\theta|\tau); \forall \tau\}$ is a class of prior distributions for θ, then the class $\mathcal{P}$ is conjugate for $\mathcal{F}$ if

$$\pi(\theta|x_{1:n}) \in \mathcal{P}, \quad \forall f(\cdot|\theta) \in \mathcal{F} \text{ and } \pi(\cdot) \in \mathcal{P}.$$

It is straightforward to specify a conjugate prior when the sampling distribution is member of the exponential family. Consider observation x_i generated from a sampling distribution in the exponential family

$$x_i|\theta \overset{\text{IID}}{\sim} \text{Ef}_k(u, g, h, \phi, \theta, c); \quad i = 1, \ldots, n$$

with density $\text{Ef}_k(x|u, g, h, \phi, \theta, c) = u(x)g(\theta) \exp(\sum_{j=1}^{k} c_j \phi_j(\theta)(\sum_{i=1}^{n} h_j(x)))$ and $g(\theta) = 1/\int u(x) \exp(\sum_{j=1}^{k} c_j \phi_j(\theta)(\sum_{i=1}^{n} h_j(x)))dx$. The likelihood function is equal to

$$f(x_{1:n}|\theta) = \prod_{i=1}^{n} u(x_i)g(\theta)^n \exp(\sum_{j=1}^{k} c_j \phi_j(\theta)(\sum_{i=1}^{n} h_j(x_i))). \tag{1.3}$$

The conjugate prior, corresponding to likelihood (1.3), admits density of the form

$$\pi(\theta|\tau) = \frac{1}{K(\tau)} g(\theta)^{\tau_0} \exp(\sum_{j=1}^{k} c_j \phi_j(\theta)\tau_j) \tag{1.4}$$

where $\tau = (\tau_0, \ldots, \tau_k)$ is such that $K(\tau) = \int_{\Theta} g(\theta)^{\tau_0} \exp(\sum_{j=1}^{k} c_j \phi_j(\theta)\tau_j)d\theta < \infty$. The resulting posterior of θ has the form

$$\pi(\theta|x_{1:n}, \tau) = \frac{1}{K(\tau^*)} g(\theta)^{\tau_0^*} \exp(\sum_{j=1}^{k} c_j \phi_j(\theta) \tau_j^*))$$

with $\tau^* = (\tau_0^*, \tau_1^*, \ldots, \tau_k^*)$, $\tau_0^* = \tau_0 + n$, and $\tau_j^* = \sum_{i=1}^{n} h_j(x_i) + \tau_j$ for $j = 1, \ldots, k$.

It is easy to see that (1.4) is conjugate to (1.3) as the posterior can be re-written as $\pi(\theta|x_{1:n}, \tau) = \pi(\theta|\tau^*)$ where $\tau^* = \tau + t_n(x_{1:n})$, and $t_n(x_{1:n}) = (n, \sum_{i=1}^{n} h_1(x_i), \ldots, \sum_{i=1}^{n} h_k(x_i))$. You can check the demo in.[2]

Example: Bernoulli model (Cont.)

Consider observations $x_{1:n} = (x_1, \ldots, x_n) \in \mathbb{R}^n$ generated from a Bernoulli distribution with success rate $\theta \in [0, 1]$; i.e., $x_i|\theta \sim \mathrm{Br}(\theta)$, $i = 1, \ldots, n$. Interest lies in specifying a conjugate prior for θ.

The sampling distribution is member of the exponential family, with $u(x) = 1$, $g(\theta) = (1 - \theta)$, $c_1 = 1$, $\phi_1(\theta) = \log(\frac{\theta}{1-\theta})$, $h_1(x) = x$, because

$$f(x|\theta) = \mathrm{Br}(x|\theta) = \theta^x (1 - \theta)^{1-x} = (1 - \theta) \exp(\log(\frac{\theta}{1 - \theta})x).$$

The corresponding conjugate prior has PDF such as

$$\pi(\theta|\tau) \propto (1 - \theta)^{\tau_0} \exp(\log(\frac{\theta}{1 - \theta})\tau_1) = \theta^{(\tau_1+1)-1}(1 - \theta)^{(\tau_0-\tau_1+1)-1},$$

where we recognize Beta distribution $\pi(\theta|\tau) = \mathrm{Be}(\theta|a, b)$, with $a = \tau_1 + 1$, $b = \tau_0 - \tau_1 + 1$. Therefore, the posterior distribution is

$$\pi(\theta|x_{1:n}, \tau) = \pi(\theta|\tau_0 + n, \tau + \sum_{i=1}^{n} h(x_i)) \propto \theta^{(\tau_1+n\bar{x}+1)-1}(1 - \theta)^{(\tau_0+n-\tau_1-n\bar{x}+1)-1}$$

which is $\mathrm{Be}(\theta|a^*, b^*)$, with $a^* = a + n\bar{x}$, and $b^* = b + n - n\bar{x}$.

1.3 Point Estimation

Often interest lies in learning the 'true' value of the unknown parameter $\theta \in \Theta$, or the future values of a future sequence of observations $y_{1:m} \in \mathcal{X}^m$; this is performed via the Bayesian point estimator. Here, we demonstrate the theory of the Bayesian point estimator in parametric inference, and leave the extension to the predictive inference to the reader.

---

[2] Web-applet: https://georgios-stats-1.shinyapps.io/demo_conjugatepriors/.

Bayes (parametric) point estimator of $\theta \in \Theta$ with respect to the loss function $\ell(\theta, \delta)$ and the posterior distribution $\pi(\theta|x_{1:n})$ is an Bayes rule δ^π which minimizes $\int_\Theta \ell(\theta, \delta)\pi(d\theta|x_{1:n})$; i.e.,

$$\delta^\pi(x_{1:n}) = \arg\min_{\forall \delta \in \Theta} E_\pi(\ell(\theta, \delta)|x_{1:n}) = \arg\min_{\forall \delta \in \Theta} \int_\Theta \ell(\theta, \delta)\pi(d\theta|x_{1:n}). \qquad (1.5)$$

Often the accuracy of the Bayes point estimator is represented by its standard error. A commonly accepted metric for the standard error of the j-th dimension of the estimator δ^π is

$$se_\pi(\delta_j|x_{1:n}) = \sqrt{MSE_\pi(\delta_j|x_{1:n})}$$

where $MSE_\pi(\delta_j|x_{1:n}) = [E_\pi((\theta - \delta)(\theta - \delta)^\top|x_{1:n})]_{j,j}$ is the mean squared error of δ_j.

A number of standard Bayesian point estimates, under different loss functions, are location summary statistics of the posterior distribution (mean, median, mode, quantiles, etc.) You can check the demo in.[3]

The Bayesian estimate of θ with respect to the linear loss $\ell(\theta, \delta) = c_1(\delta - \theta)1_{\theta \leq \delta}(\delta) + c_2(\theta - \delta)1_{\{\theta \leq \delta\}^c}(\delta)$ is the $\frac{c_2}{c_1+c_2}$-th posterior quantile; i.e., $\pi(\theta \in (-\infty, \delta(x_{1:n}))|x_{1:n}) = \frac{c_2}{c_1+c_2}$. The linear loss function essentially allows the adjustment of the penalty between over-estimating and under-estimating θ, by adjusting c_1 and c_2. In particular, for $c_1 = c_2$, we get the absolute loss $\ell(\theta, \delta) = |\theta - \delta|$ and the posterior estimator is the posterior median

$$\delta(x_{1:n}) = \text{median}_\pi(\theta|x_{1:n}). \qquad (1.6)$$

The absolute loss is more appropriate when over-estimation and under-estimation are of the same concern (as penalized the same).

The Bayes estimate $\delta^\pi(x_{1:n})$ of θ with respect to the quadratic loss function $\ell(\theta, \delta) = (\theta - \delta)^2$ is

$$\delta^\pi(x_{1:n}) = E_\pi(\theta|x_{1:n}). \qquad (1.7)$$

The posterior mean of θ as an estimator of θ essentially minimizes the estimator error $se_\pi(\delta|x_{1:n})$, which is equal to the posterior standard error. Obviously, the standard error of the estimator (1.7) is equal to the posterior standard error. Compared to the absolute loss, the quadratic loss aims at over-penalizing large but unlikely errors. In fact, quadratic loss aims at minimizing the standard error $se_\pi(\delta|x_{1:n})$.

Finally, the Bayesian estimate of θ with respect to the zero-one loss $\ell(\theta, \delta) = 1 - 1_{B_\epsilon(\delta)}(\theta)$ is the posterior mode

$$\delta(x_{1:n}) = \text{mode}_\pi(\theta|x_{1:n}) \qquad (1.8)$$

as $\epsilon \to 0$.

---

[3] Web-applet: https://georgios-stats-1.shinyapps.io/demo_PointEstimation/.

Example: Bernoulli model (Cont.)

Interest lies in calculating the Bayesian point estimator under the absolute loss function. This is the Maximum A posteriori Estimator (the posterior mode). It is

$$\log(\pi(\theta|x_{1:n})) \propto (n\bar{x} + a - 1)\log(\theta) + (n - n\bar{x} + b - 1)\log(1 - \theta).$$

For $a > 0$, $b > 0$, $\frac{d}{d\theta}\log(\pi(\theta|x_{1:n}))|_{p=\delta(x)} = 0$ implies $\delta(x) = \frac{n\bar{x}+a-1}{n+a+b-2}$. Note that (a.) if $a \to 1$, $b \to 1$ (aka $\pi(\theta|a, b) \propto 1$), then $\delta^{\pi}(x) = \bar{x}$ similar to frequentists stats; (b.) if $a \to 0$, $b \to 0$ (aka $\pi(\theta|a, b) \propto \theta^{-1}(1 - \theta)^{-1}$), then $\delta(x) = \frac{n\bar{x}-1}{n-2}$; if $a \to 1/2, b \to 1/2$ (aka $\pi(\theta|a, b) \propto \theta^{-1/2}(1 - \theta)^{-1/2}$), then $\delta(x) = \frac{n\bar{x}-1/2}{n-1}$; if $n \to \infty$, $a > 0$, $b > 0$, then $\delta(x) = \bar{x}$.

1.4   Credible Sets

Instead of just reporting parametric (or predictive) point estimates for θ (or $y_{1:m}$), it is often desirable and more useful to report a subset of values $C_a \subseteq \Theta$ (or $C_a \subseteq X^m$) where the posterior (or predictive) probability that $\theta \in C_a$ (or $y_{1:m} \in C_a$) is equal to a certain value a reflecting one's degree of believe.

The definition below describes the credible set [1, 5].

Definition 1.1 (*Posterior Credible Set*) A set $C_a \subseteq \Theta$ such that

$$\pi(\theta \in C_a|x_{1:n}) = \int_{C_a} \pi(d\theta|x_{1:n}) \geq 1 - a$$

is called '$100(1 - a)\%$' posterior credible set for θ, with respect to the posterior distribution $\pi(d\theta|x_{1:n})$.

In contrast to the frequentist stats, in Bayesian stats we can speak meaningfully of the probability that θ is in C_a, because probability $1 - a$ reflects one's degree of believe that $\theta \in C_a$.

Among all the credible sets C_a in Definition 1.1, we are often interested in those that have the minimum volume. It can be proved [2] that the highest probability density (HPD) sets have this property. HPD consider those values of θ corresponding to the highest posterior pdf/pmf (aka the most likely values of θ).

Definition 1.2 (*Posterior highest probability density (HPD) set*) The $100(1 - a)\%$ highest probability density set for $\theta \in \Theta$ with respect to the posterior distribution $\pi(\theta|x_{1:n})$ is the subset C_a of Θ of the form

$$C_a = \{\theta \in \Theta : \pi(\theta|x_{1:n}) \geq k_a\} \tag{1.9}$$

where k_a is the largest constant such that

$$\pi(\theta \in C_a | x_{1:n}) \geq 1 - a. \tag{1.10}$$

From the decision theory perspective, HPD set C_a is the Bayes estimate of C_a the credible interval under the loss function $\ell(C_a, \theta) = k|C_a| - 1_{C_a}(\theta)$, for $k > 0$ which penalizes sets with larger volumes. The proof is available in [2].

Example: Multivariate Normal model

Consider observations $x_1, \ldots, x_n$ independently drawn from a q-dimensional normal $N_q(\mu, \Sigma)$ with unknown $\mu \in \mathbb{R}^q$, $q \geq 1$, and known Σ, μ_0, Σ_0. Assume prior $\mu \sim N_q(\mu_0, \Sigma_0)$. Interest lies in calculating the C_a parametric HPD credible interval for μ.

The posterior PDF of μ is

$$\pi(\mu | x_{1:n}) \propto f(x_{1:n} | \mu) \pi(\mu) = \prod_{i=1}^{n} N_q(x_i | \mu, \Sigma) N_q(\mu | \mu_0, \Sigma_0)$$

$$\propto \exp(-\frac{1}{2}(\mu - \hat{\mu}_n)^T \hat{\Sigma}_n^{-1}(\mu - \hat{\mu}_n)) \propto N_q(\mu | \hat{\mu}_n, \hat{\Sigma}_n)$$

where $\Sigma_n = (n\Sigma^{-1} + \Sigma_0^{-1})^{-1}$, and $\hat{\mu}_n = \hat{\Sigma}_n(n\Sigma^{-1}\bar{x} + \Sigma_0^{-1}\mu_0)$. So $\mu | x_{1:n} \sim N_q(\hat{\mu}_n, \hat{\Sigma}_n)$.

From Definition 1.2, the credible set has the form

$$C_a = \{\mu \in \mathbb{R}^q : \pi(\mu | x_{1:n}) \geq k_a\}$$
$$= \{\mu \in \mathbb{R}^q : (\mu - \hat{\mu}_n)^T \hat{\Sigma}_n^{-1}(\mu - \hat{\mu}_n) \leq -\log(2\pi \det(\hat{\Sigma}_n)))k_a = \tilde{k}_a\}$$

where k_a is the greatest value satisfying

$$\pi_{N_q(\hat{\mu}_n, \hat{\Sigma}_n)}(\mu \in C_a | x_{1:n}) \geq 1 - a \iff$$
$$\pi_{\chi_q^2}((\mu - \hat{\mu}_n)^T \hat{\Sigma}_n^{-1}(\mu - \hat{\mu}_n) \leq \tilde{k}_a) \geq 1 - a. \tag{1.11}$$

Here, $(\mu - \hat{\mu}_n)^T \hat{\Sigma}_n^{-1}(\mu - \hat{\mu}_n) \sim \chi_q^2$ as a sum of squares of independent standard normal random variables, and hence $\tilde{k}_a$ is the $1 - a$-th quantile of the χ_q^2 distribution; i.e., $\tilde{k}_a = \chi_{q,1-a}^2$. Therefore, C_a parametric HPD credible set for μ is

$$C_a = \{\mu \in \mathbb{R}^q : (\mu - \hat{\mu}_n)^T \hat{\Sigma}_n^{-1}(\mu - \hat{\mu}_n) \leq \chi_{q,1-a}^2\}$$

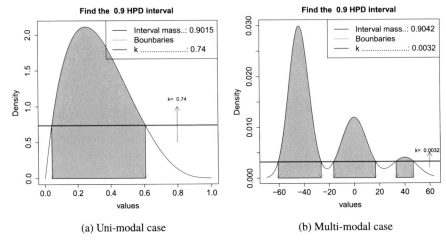

(a) Uni-modal case   (b) Multi-modal case

Fig. 1.1 Schematic of 1D HPD set

In real applications, the calculation of the credible interval might be intractable, due to the inversion in (1.9) or integration in (1.10). Below, we present a Naive algorithm [1] that can be implemented in a computer.[4]

- Create a routine which computes all solutions θ^* to the equation $\pi(\theta|x_{1:n}) = k_a$, for a given k_a. Typically, $C_a = \{\theta \in \Theta : \pi(\theta|x_{1:n}) \geq k_a\}$ can be constructed from those solutions.
- Create a routine which computes $\pi(\theta \in C_a|x_{1:n}) = \int_{\theta \in C_a} \pi(\theta|x_{1:n})d\theta$
- Numerically solve the equation $\pi(\theta \in C_a|x_{1:n}) = 1 - a$ as k_a varies.

Figure 1.1 demonstrates the above procedure in 1D unimodal and tri-modal cases. Specifically, the red horizontal bar denotes k_a moves upwards, and intersects the density at locations which are the potential boundaries of C_a. The bar stops to move when the total density above regions of the parametric space is equal to $1 - \alpha$. The HPD credible set results as the union of these sub-regions. You can check the demo in.[5]

Theorem 1.1 suggests a computationally convenient way to calculate HPD credible intervals in 1D, and unimodal cases. The proof is available in [3].

Theorem 1.1 *Let θ follows a distribution with unimodal density $\pi(\theta|x_{1:n})$. If the interval $C_a = [L, U]$ satisfies*

1. $\int_L^U \pi(\theta|x_{1:n})d\theta = 1 - a$,
2. $\pi(U) = \pi(L) > 0$, and
3. $\theta_{mode} \in (L, U)$, where θ_{mode} is the mode of $\pi(\theta|x_{1:n})$,

then interval $C_a = [L, U]$ is the HPD interval of θ with respect to $\pi(\theta|x_{1:n})$.

---

[4] Web-applet: https://georgios-stats-1.shinyapps.io/demo_crediblesets/.
[5] Web-applet: https://georgios-stats-1.shinyapps.io/demo_crediblesets/.

Example: Bernoulli model (Cont.)

Interest lies in calculating the 2-sides 95% HPD interval for θ, given a sample with $n = 30$, and $\sum_{i=1}^{30} x_i = 15$, and prior hyper-parameters $a = b = 2$.

The posterior distribution of θ is $\text{Be}(a + n\bar{x} = 17, b + n - n\bar{x} = 17)$, which is 1D and unimodal; hence we use Theorem 1.1. It is

$$1 - a = \int_L^U \text{Be}(\theta | 17, 17) d\theta = \text{Be}(\theta < U | 17, 17) - \text{Be}(p < L | 17, 17).$$

Note that Beta PDF is symmetric around 0.5 when $a^* = b^*$, and so is here where $\text{Be}(17, 17)$. Then,

$$1 - a = \text{Be}(\theta < U | 17, 17) - (1 - \text{Be}(\theta < U | 17, 17)) = 2\text{Be}(\theta < U | 17, 17) - 1$$

so $\text{Be}(\theta < U | 17, 17) = 1 - a/2$ and $L = 1 - U$. For $a = 0.95$, the 95% posterior credible interval for θ is $[L, U] = [0.36, 0.64]$.

Remark 1.1 Predictive credible sets for a future sequence of observations $y_{1:m}$, are defined and constructed as parametric ones by replacing θ with $y_{1:m}$ and $\pi(x_{1:n}|\theta)$ with $p(y_{1:m}|x_{1:n})$ in Definitions 1.1 and 1.2, and their consequences in this section. It is left as an Exercise.

1.5 Hypothesis Test

Often there is interest in reducing the overall parametric space Θ (aka the set of possible values of that the uncertain parameter θ can take) to a smaller subset. For instance; whether the proportion of Brexiters is larger than 0.5 ($p > 0.5$) or not ($p \leq 0.5$).

Such a decision can be formulated as a hypothesis test [1], namely the decision procedure of choosing between two non-overlapping hypotheses

$$H_0 : \theta \in \Theta_0 \quad \text{vs} \quad H_1 : \theta \in \Theta_1 \tag{1.12}$$

where $\{\Theta_0, \Theta_1\}$ partitions the space Θ. Typically, hypotheses, $\{H_k\}$, are categorized in three categories. Single hypothesis for θ is called the hypothesis where $\Theta_j = \{\theta_j\}$ contains a single element. Composite hypothesis for θ is called the hypothesis where $\Theta_j \subseteq \Theta$ contains many elements. General alternative hypothesis for θ is called the composite hypothesis where $\Theta_1 = \Theta - \{\theta_0\}$ when it is compared against a single hypothesis $H_0 : \theta = \theta_0$. It is denoted as $H_1 : \theta \neq \theta_0$.

Based on the partitioning implied by (1.12), the overall prior π can be expressed as $\pi(\theta) = \pi_0 \times \pi_0(\theta) + \pi_1 \times \pi_1(\theta)$ where $\pi_k = \int_{\Theta_k} \pi(d\theta)$, and $\pi_k(\theta) = \frac{\pi(\theta)1_{\Theta_k}(\theta)}{\int_{\Theta_k} \pi(d\theta)}$.

Here, π_0, and π_1 describe the prior probabilities on H_0 and H_1, respectively, while $\pi_0(\theta)$ and $\pi_1(\theta)$ describe how the prior mass is spread out over the hypotheses H_0 and H_1, respectively.

We could see the hypothesis testing (1.12) as parametric point inference about the indicator function

$$1_{\Theta_1}(\theta) = \begin{cases} 0 & , \theta \in \Theta_0 \\ 1 & , \theta \in \Theta_1 \end{cases}. \tag{1.13}$$

To estimate (1.13), a reasonable loss function $\ell(\theta, \delta)$ would be the $c_\mathrm{I} - c_\mathrm{II}$ loss function

$$\ell(\theta, \delta) = \begin{cases} 0 & , \text{if } \theta \in \Theta_0, \delta = 0 \\ 0 & , \text{if } \theta \notin \Theta_0, \delta = 1 \\ c_\mathrm{II} & , \text{if } \theta \notin \Theta_0, \delta = 0 \\ c_\mathrm{I} & , \text{if } \theta \in \Theta_0, \delta = 1 \end{cases} \tag{1.14}$$

where $c_\mathrm{I} > 0$ and $c_\mathrm{II} > 0$ are specified by the researcher. Here, $c_\mathrm{I} > 0$ (and $c_\mathrm{II} > 0$) denote the loss if we decide to accept H_0 (and H_1) while the correct answer would be to choose H_1 (H_0). According to (1.5), under (1.14), the Bayes estimator of (1.13) is

$$\delta(x_{1:n}) = \begin{cases} 0 & , \text{if } \pi(\theta \in \Theta_0 | x_{1:n}) > \frac{c_\mathrm{II}}{c_\mathrm{II} + c_\mathrm{I}} \\ 1 & , \text{otherwise} \end{cases} \tag{1.15}$$

where $\pi(\theta \in \Theta_0 | x_{1:n}) = \int_{\Theta_0} \pi(d\theta | x_{1:n})$. In other words, hypothesis H_1 is accepted if $\frac{\pi(\theta \in \Theta_0 | x_{1:n})}{\pi(\theta \in \Theta_1 | x_{1:n})} < \frac{c_\mathrm{II}}{c_\mathrm{I}}$.

Hypothesis tests in Bayesian statistics can also be addressed with the aid of Bayes factors. Bayes factor $B_{01}(x_{1:n})$ is the ratio of the posterior probabilities of H_0 and H_1 over the ratio of the prior probabilities of H_0 and H_1

$$B_{01}(x_{1:n}) = \frac{\pi(\theta \in \Theta_0 | x_{1:n})/\pi(\theta \in \Theta_0)}{\pi(\theta \in \Theta_1 | x_{1:n})/\pi(\theta \in \Theta_1)} \tag{1.16}$$

$$= \begin{cases} \frac{f(x_{1:n}|\theta_0)}{f(x_{1:n}|\theta_1)} & ; H_0 : \text{single vs } H_1 : \text{single} \\ \frac{\int_{\Theta_0} f(x_{1:n}|\theta)\pi_0(d\theta)}{\int_{\Theta_1} f(x_{1:n}|\theta)\pi_1(d\theta)} & ; H_0 : \text{composite vs } H_1 : \text{composite} \\ \frac{f(x_{1:n}|\theta_0)}{\int_{\Theta_1} f(x_{1:n}|\theta)\pi_1(d\theta)} & ; H_0 : \text{single vs } H_1 : \text{composite} \end{cases}. \tag{1.17}$$

Under the $c_\mathrm{I} - c_\mathrm{II}$ loss function, (1.15) implies that one would accept H_0 if $B_{01}(x_{1:n}) > \frac{c_\mathrm{II}}{c_\mathrm{I}} \frac{\pi_1}{\pi_0}$, and accept H_1 if otherwise. Alternatively, Jeffreys [6] developed a scale rule (Table 1.1) to judge the strength of evidence in favor of H_0 or against H_0 brought by the data. Although Jeffreys' rule avoids the need to specify c_I and c_{II}, it is a heuristic rule-of-thumb guide, not based on decision theory concepts, and hence many researchers argue against its use.

Table 1.1 Jeffreys' scale rule [6]

B_{01}	$\log_{10}(B_{01})$	Strength of evidence
$(1, +\infty)$	$(0, +\infty)$	H_0 is supported
$(10^{-1/2}, 1)$	$(-1/2, 0)$	Evidence against H_0: not worth more than a bare
$(10^{-1}, 10^{-1/2})$	$(-1, -1/2)$	Evidence against H_0: substantial
$(10^{-3/2}, 10^{-1})$	$(-3/2, -1)$	Evidence against H_0: strong
$(10^{-2}, 10^{-3/2})$	$(-2, -3/2)$	Evidence against H_0: very strong
$(0, 10^{-2})$	$(-\infty, -2)$	Evidence against H_0: decisive

Example: Bernoulli model (Cont.)

We are interested in testing the hypotheses $H_0 : \theta = 0.5$ and $H_1 : \theta \neq 0.5$, given that $\pi_0 = 1/2$, and using the $c_I - c_{II}$ loss function with $c_I = c_{II}$. Here, $\Theta_0 = \{0.5\}$ and $\Theta_1 = [0, 0.5) \cup (0.5, 1]$. The overall prior is $\pi(\theta) = \pi_0 1_{\theta_0}(\theta) + (1 - \pi_0)\text{Be}(\theta | a, b)$. The Bayes factor is

$$B_{01}(x_{1:n}) = \frac{\prod_{i=1}^{n} \text{Br}(x_i | \theta_0)}{\int_{(0,1)} \prod_{i=1}^{n} \text{Br}(x_i | \theta)\text{Be}(\theta | a, b)d\theta} = \frac{\theta_0^{x_*}(1 - \theta_0)^{n-x_*}}{\text{B}(n\bar{x} + a, n - n\bar{x} + b)/\text{B}(a, b)}.$$

Given $a = b = 2$, $n = 30$, and $\sum_{i=1}^{30} x_i = 15$, it is $B_{01}(x_{1:n}) = 18.47 > c_{II}/c_I = 1$. Hence, we accept H_1.

1.5.1 Model Selection

Often the researcher is uncertain which statistical model (sampling distribution) can better represent the real data generating process. There is a set $\mathcal{M} = \{m_1, m_2, \ldots\}$ of candidate statistical models $m_k = \{f_k(\cdot | \varphi_k); \varphi_k \in \Phi_k\}$, where $f_k(\cdot | \varphi_k)$ denotes the sampling distribution, and φ_k denotes the unknown parameters for $k = 1, 2, \ldots$ Let $\pi_k = \pi(m_k)$ denote the marginal model prior and $\pi_k(\varphi_k) = \pi(\varphi_k | m_k)$ denote the prior of the unknown parameters φ_k of given model m_k.

Selection of the 'best' model from a set of available candidate models can be addressed via hypothesis testing. For simplicity, we consider there are only two models m_0 and m_1 with unknown parameters $\vartheta_0 \in \Phi_0$ and $\vartheta_1 \in \Phi_1$. Then, model selection is performed as a hypothesis test

$$H_0 : (m, \varphi) \in \Theta_0 \quad \text{vs} \quad H_1 : (m, \varphi) \in \Theta_1 \tag{1.18}$$

where $\Theta_k = \{m_k\} \times \Phi_k$, $\Theta = \cup_k \Theta_k$. The overall joint prior is specified as $\pi(m, \varphi) = \pi_0 \times \pi_0(\varphi_0) + \pi_1 \times \pi_1(\varphi_1)$ on $(m, \varphi) \in \Theta$ where $\Theta = \cup_k \Theta_k$, where $\pi_k(\varphi_k) = \frac{\pi(m,\varphi) 1_{m_k}(m)}{\int_{\Phi_k} \pi(m,d\varphi)}$ on $\varphi_k \in \Phi_k$, and $\pi_k = \int_{\Theta_k} \pi(m_k, d\varphi_k)$. Now the model selection problem has been translated into a hypothesis test.

Example: Negative binomial vs. Poisson model [2]

We are interested in testing the hypotheses

$$H_0 : x_i|\phi \sim Nb(\phi, 1), \quad \phi > 0, \quad \text{vs.} \quad H_1 : x_i|\lambda \sim Pn(\lambda), \quad \lambda > 0$$

by using the $c_I - c_{II}$ loss function with $c_I = c_{II}$. Consider two observations $x_1 = x_2 = 2$ are available. Consider overall prior $\pi(\theta)$ with density $\pi(\theta) = \pi_0 Be(\phi|a_0, b_0) + \pi_1 Ga(\lambda|a_1, b_1)$ with $\pi_0 = \pi_1 = 0.5$.

This is a composite vs. composite hypothesis test. It is

$$\int_{\Theta_0} f(x_{1:n}|\varphi_0)\pi_0(d\varphi_0) = \frac{\Gamma(a_0 + b_0)}{\Gamma(a_0)\Gamma(b_0)} \int_0^1 \phi^{n+a_0-1}(1-\phi)^{n\bar{x}+b_0-1}d\phi$$

$$= \frac{\Gamma(a_0 + b_0)}{\Gamma(a_0)\Gamma(b_0)} \frac{\Gamma(n + a_0)\Gamma(n\bar{x} + b_0)}{\Gamma(n + n\bar{x} + a_0 + b_0)}$$

$$\int_{\Theta_1} f(x_{1:n}|\varphi_1)\pi_1(d\varphi_1) = \frac{b_1^{a_1}}{\Gamma(a_1)(n + b_1)^{n\bar{x}+a_1}} \int_0^\infty \lambda^{n\bar{x}+a_1-1} \exp(-(n + b_1)\lambda)d\lambda$$

$$= \frac{\Gamma(n\bar{x} + a_1)}{\Gamma(a_1)(n + b_1)^{n\bar{x}+a_1}} \frac{1}{\prod_{i=1}^n x_i!}$$

and hence $B_{01}(x_{1:n}) = \frac{\Gamma(a_0+b_0)}{\Gamma(a_0)\Gamma(b_0)} \frac{\Gamma(n+a_0)\Gamma(n\bar{x}+b_0)}{\Gamma(n+n\bar{x}+a_0+b_0)} \frac{\Gamma(a_1)(n+b_1)^{n\bar{x}+a_1}}{\Gamma(n\bar{x}+a_1)} \prod_{i=1}^n x_i!$. It is $B_{01}(x_{1:n}) = 0.29 > 1$, and hence I accept H_1 and the Poisson model.

References

1. James O Berger. *Statistical decision theory and Bayesian analysis*. Springer Science & Business Media, 2013.
2. José M Bernardo and Adrian FM Smith. *Bayesian theory*, volume 405. John Wiley & Sons, 2009.
3. George Casella and Roger L Berger. *Statistical inference*, volume 2. Duxbury Pacific Grove, CA, 2002.
4. Bruno De Finetti. *Theory of probability: a critical introductory treatment*, volume 6. John Wiley & Sons, 2017.

5. Morris H DeGroot. *Optimal statistical decisions*, volume 82. John Wiley & Sons, 2005.
6. Harold Jeffreys. *The theory of probability*. OUP Oxford, 1998.

Chapter 2
Sampling from Complex Probability Distributions: A Monte Carlo Primer for Engineers

Louis J. M. Aslett

Abstract Models which are constructed to represent the uncertainty arising in engineered systems can often be quite complex to ensure they provide a reasonably faithful reflection of the real-world system. As a result, even computation of simple expectations, event probabilities, variances, or integration over utilities for a decision problem can be analytically intractable. Indeed, such models are often sufficiently high dimensional that even traditional numerical methods perform poorly. However, access to random samples drawn from the probability model under study typically simplifies such problems substantially. The methodologies to generate and use such samples fall under the stable of techniques usually referred to as 'Monte Carlo methods'. This chapter provides a motivation, simple primer introduction to the basics, and sign-posts to further reading and literature on Monte Carlo methods, in a manner that should be accessible to those with an engineering mathematics background. There is deliberately informal mathematical presentation which avoids measure-theoretic formalism. The accompanying lecture can be viewed at https://www.louisaslett.com/Courses/UTOPIAE/.

2.1 Motivation

There is a natural tension when constructing a probabilistic model with the aim of encapsulating the uncertainty in an engineered system: on the one hand, there is a desire to capture every nuance of the system to fully reflect all knowledge about its behaviour; on the other, there is a drive towards parsimony for reasons of interpretability, robustness, and computability. Interpretability and robustness are important goals and should indeed guide a reduction in model complexity, but reducing model complexity purely to enable computability would seem a hinderance, especially if that parsimony impedes answering the research questions at hand since, put simply, 'reality can be complicated' [7]. As such, the methodology of this chapter

L. J. M. Aslett (✉)
Department of Mathematical Sciences, Durham University, Durham, United Kingdom
e-mail: louis.aslett@durham.ac.uk

15

should not be employed simply to enable an inappropriately complex model, but rather serves to facilitate the use of models which are complex enough when judged by purely subject matter and statistical concerns.

Monte Carlo methods have played a crucial role in a vast array of applications of statistical methodology, from the prediction of future marine species discoveries [29] through to reconstruction of the ancient climate on Earth [16]; from criminal justice offending risk [17] to inferring networks of corporate governance through the financial crash [12]; and from estimating bounds on engineering system survival functions [11] to the assessment of offshore oil production availability [30]. The utility of Monte Carlo in these applications varies substantially, from estimation of confidence intervals and event probabilities, through optimisation methods to full evaluation of Bayesian posterior distributions for parameter inference.

With this breadth of application in mind, we may assume hereinafter that we have a probabilistic model for some engineered system of interest which—after considering all subject matter and statistical concerns—is too complex to be able to compute relevant quantities of interest (be they event probabilities, confidence intervals, posterior distributions, etc.). As a concrete example, if one were to construct a Bayesian model of reliability using ideas introduced in Chap. 1, then our model would comprise some prior distribution over the vector of model parameters, $\pi(\theta)$, together with a generative model for the failure time depending on those parameters, $\pi(t \mid \theta)$. After collecting some lifetime data $\underline{t} = \{t_1, \ldots, t_n\}$, the most simple research question of interest may be the posterior expected value of the parameters:

$$\mathbb{E}_\pi[\theta] = \int_\Omega \theta \, \pi(\theta \mid \underline{t}) \, d\theta = \frac{1}{c} \int_\Omega \theta \, \pi(\theta) \prod_{i=1}^n \pi(t_i \mid \theta) \, d\theta \qquad (2.1)$$

where Ω is the space of all possible parameter values and c is a normalising constant.

Indeed, it is traditional in Monte Carlo literature to focus attention on the computation of expectations with respect to some probability density under consideration, which need not necessarily be a Bayesian posterior. That is, given a general probability model $\pi(x), x \in \Omega$, and a functional $f : \Omega \to \mathbb{R}$, interest is typically in:

$$\mathbb{E}_\pi[f(X)] := \int_\Omega f(x)\pi(x) \, dx \qquad (2.2)$$

and this is the perspective that will be adopted in this chapter.

We complete our motivation of Monte Carlo in this Section by highlighting the generality of expectations of the form (2.2), followed by a short discussion of standard numerical integration techniques. In Sect. 2.2, the Monte Carlo estimator and its error analysis are recapped and contrasted with numerical integration. The core methods of Monte Carlo simulation are introduced in Sect. 2.3, with pointers to more advanced material in Sect. 2.4. Note that we will in places abuse formal notation where we believe it aids intuitive understanding since the goal of this chapter is to be a basic primer, not a rigorous treatment.[1] A first course in probability and statistics are assumed background.

The accompanying lecture from the UTOPIAE training school can be viewed at https://www.louisaslett.com/Courses/UTOPIAE/.

2.1.1 Generality of Expectations

The formulation in (2.2) may appear rather restrictive to the uninitiated reader. However, considering only expectations of this form does not result in any loss of generality. For example, (re-)defining:

$$\pi(x) := \frac{1}{c}\pi(x)\prod_{i=1}^{n}\pi(t_i \mid x)$$

$$f(x) := x$$

means that (2.2) simply becomes the posterior expectation in (2.1). However, one should note that arbitrary statements of probability are also computable as expectations. That is,

$$\mathbb{P}(X < a) = \int_{-\infty}^{a}\pi(x)\,dx = \int_{\Omega}\mathbb{I}_{(-\infty,a]}(x)\pi(x)\,dx = \mathbb{E}_{\pi}[\mathbb{I}_{(-\infty,a]}(X)]$$

where for a general set $E \subseteq \Omega$,

$$\mathbb{I}_E(x) := \begin{cases} 1 & \text{if } x \in E \\ 0 & \text{if } x \notin E \end{cases}$$

That is, to evaluate the probability of an arbitrary event, $\mathbb{P}(X \in E)$, simply set $f(X) := \mathbb{I}_E(X)$ when evaluating (2.2).

2.1.2 Why Consider Monte Carlo?

In some special cases, the integral (2.2) may have an analytical solution and in such situations one should not resort to Monte Carlo or other methods. When there is no known analytical form for the integral, a reader with a general mathematical

---

[1] For example, we will write '$\mathbb{P}(X = x)$' even where X is continuous to emphasise the link to the density function and will use $\pi(x)$ to reference both a target distribution or prior where the meaning is clear from context. For the more advanced reader there are already many excellent more rigorous treatments in the literature, some of which we reference towards the end.

background may be tempted to reach for a numerical integration method, such as a simple mid-point Riemann integral or a more sophisticated quadrature approach.

Consider the mid-point Riemann integral in the simple 1-dimensional setting. Letting $g(x) := f(x)\pi(x)$, then the expectation would be approximated using n evaluations by:

$$\int_a^b g(x)\,dx \approx \frac{b-a}{n}\sum_{j=1}^{n} g(x_j), \tag{2.3}$$

where

$$x_j := a + \frac{b-a}{n}\left(j - \frac{1}{2}\right).$$

The absolute error in using (2.3) is bounded [24, Theorem 7.1]:

$$\left| \int_a^b g(x)\,dx - \frac{b-a}{n}\sum_{j=1}^{n} g(x_j) \right| \le \frac{(b-a)^3}{24n^2} \max_{a \le z \le b} |g''(z)|.$$

Clearly, $\frac{(b-a)^3}{24} \max_{a \le z \le b}|g''(z)|$ is fixed by the problem at hand and cannot be altered by the engineer, so we achieve the accuracy we require by controlling n^{-2}—that is, by using a finer grid to compute the integral. As such, we say the error in the mid-point Riemann integral in 1 dimension is $O\left(n^{-2}\right)$—that is, if double the computational effort is expended by computing on a grid of twice as many points $(2n)$, then the worst case error is reduced by a factor of 4. This fast reduction in error and an explicit bound on it are very attractive properties.

However, as the dimension of x increases, the Riemann integral's effectiveness diminishes substantially. In general, the error of mid-point Riemann integration in d-dimensions is $O\left(n^{-2/d}\right)$. For example, even in a modest 10-dimensional problem, when the computational effort is doubled the worst case error is only reduced by a factor of ≈ 1.15. Put another way, to halve the worst case error in a 10-dimensional problem requires $\exp\left(\frac{10}{2}\log 2\right) = 32$ times the computational effort. This problem has been coined the 'curse of dimensionality'.

Of course, the Riemann integral is not the best numerical integration method, but even Simpson's rule only improves this to $O\left(n^{-4/d}\right)$. In general Bakhvalov's Theorem bounds all possible quadrature methods by $O(n^{-r/d})$, where r is the number of continuous derivatives of $g(\cdot)$ which exist and are exploited by the quadrature rule [24].

The striking result which motivates the study of Monte Carlo methods is that for a d-dimensional problem, the (mean-square)^2 error is $O\left(n^{-1/2}\right)$. The most important point to note is the absence of d in the order of the error: increasing the computational effort by some fixed amount has the same relative effect on the worst case error regardless of dimension. Of course, the devil in the detail is that the constant

---

^{2} Note that randomised simulation methods such as Monte Carlo typically report mean-square error rather than absolute error bounds.

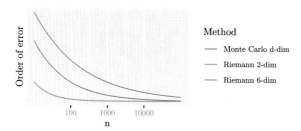

Fig. 2.1 The order of error reduction—that is, only the leading $O\left(n^{f(d)}\right)$ term—is plotted against different computational effort n. Note that all these curves would be multiplied by a different (fixed) problem dependent constant

factor which we are ignoring in that statement almost certainly has some dimension dependence, but this is true for quadrature methods too. Figure 2.1 illustrates the differences.

Consequently, Monte Carlo methods are well suited to address the problem of analysing complex probabilistic models of engineered systems, since this is precisely a setting where the parameter dimension is likely to be large.

2.2 Monte Carlo Estimators

The standard Monte Carlo estimator of the integral (2.2) is

$$\mu \triangleq \int_\Omega f(x)\pi(x)\,dx \approx \frac{1}{n}\sum_{j=1}^n f(x_j) \triangleq \hat{\mu}, \tag{2.4}$$

where $x_j \sim \pi(\cdot)$. In other words, the problem of integration is transformed instead into the problem of drawing random samples x_j distributed according to the probability density $\pi(\cdot)$. Importantly, this estimator is unbiased, that is, $\mathbb{E}[\hat{\mu}] = \mu$.

If the samples x_j are independently and identically distributed (iid) according to $\pi(\cdot)$, then the root mean-square error of the estimator $\hat{\mu}$ is

$$\text{RMSE} := \sqrt{\mathbb{E}_\pi\left[\left(\int f(x)\pi(x)\,dx - \frac{1}{n}\sum_{j=1}^n f(x_j)\right)^2\right]} = \frac{\sigma}{\sqrt{n}},$$

where $\sigma^2 = \mathrm{Var}_\pi (f(X))$. Again, part of this error is (mostly) inherent to the problem[3]—σ in this case—so that we achieve desired accuracy by controlling $n^{-1/2}$. There are at least three very attractive conclusions we can draw from this form:

1. as mentioned already, the relative error reduction achieved by additional computational effort is independent of dimension;
2. there is no explicit dependence on how smooth the functional, $f(\cdot)$, or probability density, $\pi(\cdot)$, are (though these may influence σ);
3. in contrast to quadrature methods, an estimate of the error can be computed from the work already done to compute the integral, by computing the empirical standard deviation of the functional of the samples drawn from $\pi(\cdot)$.

Although an absolute error is not available for a randomised method like this, a simple application of Chebyshev's inequality does provide a probabilistic bound on the absolute error exceeding a desired tolerance:

$$\mathbb{P}(|\hat{\mu} - \mu| \geq \varepsilon) \leq \frac{\mathbb{E}_\pi[(\hat{\mu} - \mu)^2]}{\varepsilon^2} = \frac{\sigma^2}{n\varepsilon^2}.$$

Indeed, it is also possible to invoke the iid Central Limit Theorem so that asymptotically,

$$\mathbb{P}\left(\frac{\hat{\mu} - \mu}{\sigma n^{-1/2}} \leq z\right) \xrightarrow{n \to \infty} \Phi(z),$$

where $\Phi(z)$ denotes the standard Normal cumulative distribution function (CDF). This enables the formation of confidence intervals for μ based on large n samples.

The discussion to date has tacitly assumed that simulating from arbitrary probability distributions $\pi(\cdot)$ is possible and relatively efficient. In fact, most Monte Carlo research is devoted to this effort since, as touched on above, there is rich and well-established theory when such samples are available. Therefore, for the remainder of this chapter, our attention turns away from discussion of the integrals which are of primary interest and focuses on the problem of simulating from arbitrary probability distributions $\pi(\cdot)$. Once these samples are available, the results above can be used to analyse the resulting estimators.

2.3 Simple Monte Carlo Sampling Methods

In this section we introduce some simple Monte Carlo methods which enable sampling from a wide array of probability distributions. Note that understanding these simple methods is crucial as they are extensively used as building blocks of more sophisticated sampling methodology.

---

[3] There are advanced Monte Carlo methods which can reduce this variance, but this is beyond the scope of this chapter. See for example [24, Chap. 8].

Almost all Monte Carlo procedures start from the assumption that we have available an unlimited stream of iid uniformly distributed values, typically on the interval $[0, 1] \subset \mathbb{R}$. How to generate such an iid stream is beyond the scope of this introductory chapter, but the interested reader may consult [13, Chaps. 1–3] and [21]. Arguably the current gold standard algorithm remains that in [22]. Typically, the average user of Monte Carlo need not worry about such issues and may rely on the high quality generators built into software such as R [26].

Thus the objective hereinafter is to study how to convert a stream $u_i \sim \text{Unif}(0, 1)$ into a stream $x_j \sim \pi(\cdot)$, where x_j is generated by some algorithm depending on the stream of u_i. In more advanced methods (see MCMC), x_j may also depend on x_{j-1} or even $x_1, \ldots, x_{j-1}$.

2.3.1 Inverse Sampling

Arguably the simplest example of generating non-uniform random variates is inverse sampling, which typically applies only to 1-dimensional probability distributions (though higher dimensional extensions have been studied). Let $F(x) := \mathbb{P}(X \leq x)$ be the cumulative distribution function (CDF) for the target probability density function $\pi(\cdot)$. Then, inverse sampling requires the inverse of the cdf, $F^{-1}(\cdot)$, which is then applied to a uniform random draw. Precisely, see Algorithm 2.1.

Algorithm 2.1 Inverse sampling algorithm

1: **procedure** INVERSE SAMPLING($F^{-1}(\cdot)$) ▷ Generate random sample from distribution with inverse CDF $F^{-1}(\cdot)$
2:    $u \sim \text{Unif}(0, 1)$
3:    $x \leftarrow F^{-1}(u)$
4:    **return** x
5: **end procedure**

To prove that the sample returned by Algorithm 2.1 is distributed according to $\pi(\cdot)$ is straight-forward. We do so by computing the CDF, $\mathbb{P}(X \leq x)$, of the X generated by this algorithm and show that this agrees with the CDF of $\pi(\cdot)$. The first step substitutes $X = F^{-1}(U)$, where $U \sim \text{Unif}(0, 1)$, as per the algorithm:

$$
\begin{aligned}
\mathbb{P}(X \leq x) &= \mathbb{P}(F^{-1}(U) \leq x) \\
&= \mathbb{P}(F(F^{-1}(U)) \leq F(x)) \qquad \text{applying } F(\cdot) \text{ to both sides} \\
&= \mathbb{P}(U \leq F(x)) \qquad\qquad \text{Uniform CDF } \mathbb{P}(U \leq u) = u \\
&= F(x).
\end{aligned}
$$

Note that applying $F(\cdot)$ to both sides in the second line is valid, since the cumulative distribution function is a non-decreasing function by definition.

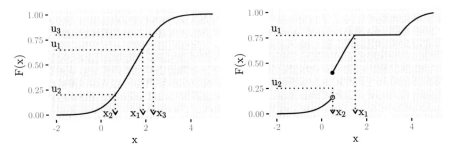

Fig. 2.2 Inverse sampling for both a continuous distribution function (left) and one containing jump discontinuities and regions of zero probability (right). Uniform random draws u are sampled and inverted through the distribution function in the obvious way (left), or by taking the infimum over values of x such that $F(x) \geq u$ (right). In the right illustration, the hypothetical ('Hypothetical' since strictly speaking this is an event of probability zero) u_1 coincides with the value at which $F(x)$ is constant and u_2 lies within the jump discontinuity

One subtlety to be aware of is that for discrete distributions or continuous distributions with jump discontinuities or areas of no support, we must define:

$$F^{-1}(u) = \inf\{x : F(x) \geq u\}, \ \forall \, u \in [0, 1].$$

It may be tempting when $F^{-1}(\cdot)$ is not available to use a numerical solver to solve $F(x) = u$ in place of line 3 in Algorithm 2.1. However, caution is required since this can result in bias [10, p. 31]. The procedure of inverse sampling is illustrated in Fig. 2.2.

Notice that this is univariate, yet earlier we saw that numerical integration will give better error bounds than Monte Carlo for low dimensional problems—as such one may choose not to use inverse sampling to actually evaluate univariate expectations. However, we often need a set of random draws from non-uniform univariate distributions which feed into a broader Monte Carlo algorithm, which is itself sampling in higher dimensions: in such situations inverse sampling is very useful. Indeed, if you use the `rnorm` function in R [26], it has used inverse sampling to generate random draws from the Normal distribution since 2003 (see `/src/nmath/snorm.c` lines 265–270), prior to that using [18] since at least v0.62 in 1998.

A final comment: inverse sampling is a special case of general transformation sampling. If one can generate samples from one distribution, there may be an appropriate transformation to turn these into samples from another distribution that may be more tractable or faster than inverse sampling. For further details, see for example [24, Chap. 4.6].

2.3.1.1 Example

In order to use inversion sampling for a Weibull distribution with shape k and scale σ, $X \sim \text{Weibull}(k, \sigma)$, we note that

$$\pi(x) = \frac{k}{\sigma}\left(\frac{x}{\sigma}\right)^{k-1} e^{-\left(\frac{x}{\sigma}\right)^k}, \quad x \in [0, \infty), \sigma > 0, k > 0$$

$$F(x) = 1 - \exp\left\{-\left(\frac{x}{\sigma}\right)^k\right\}.$$

To find $F^{-1}(u)$, set $1 - \exp\left\{-\left(\frac{x}{\sigma}\right)^k\right\} = u$ and solve for x:

$$\implies x = F^{-1}(u) = \sigma\left(-\log(1 - u)\right)^{1/k} \sim \pi(\cdot). \tag{2.5}$$

In order to generate samples from the Weibull we, therefore, take values, u, from a Uniform random number stream and transform them using (2.5).

2.3.2 Rejection Sampling

Our first higher dimensional method is an elegant algorithm, which actually crops up in more advanced guises at the cutting edge of modern Monte Carlo methods (e.g. [9, 25]). Here, the goal is to find another distribution, say $\tilde{\pi}(\cdot)$, which is easier to sample from (perhaps even using inverse sampling) and where we can construct a bound on the density function:

$$\pi(x) \leq c\tilde{\pi}(x) \quad \forall x \in \Omega, \tag{2.6}$$

where $c < \infty$ and where π and $\tilde{\pi}$ need not be normalised probability densities. We call $\tilde{\pi}(\cdot)$ the 'proposal' density, since samples will be drawn from this and then exactly the correct proportion of them retained in order to end up with a stream of samples from $\pi(\cdot)$. The full procedure is detailed in Algorithm 2.2.

We will proceed based on the assumption that $\pi(\cdot)$ and $\tilde{\pi}(\cdot)$ are normalised densities. However, note that the algorithm is also valid for un-normalised densities, so long as there still exists a c satisfying (2.6) for the un-normalised densities.

The efficiency of the algorithm hinges entirely on the value of c, so that it should be chosen as small as possible. This is because, letting A be the random variable for acceptance of a proposed sample X, the acceptance probability is (abusing notation to aid intuition):

Algorithm 2.2 Rejection sampling algorithm

1: **procedure** REJECTION SAMPLING($\pi(\cdot)$, $\tilde{\pi}(\cdot)$, c)   ▷ Generate random sample from distri-
                                                                                                          bution with unnormalised density $\pi(\cdot)$

2:   $a \leftarrow$ FALSE
3:   **while** $a =$ FALSE **do**   ▷ Repeat until acceptance
4:       $u \sim$ Unif$(0, 1)$
5:       $x \sim \tilde{\pi}(\cdot)$ ▷ Propose a possible sample
6:       **if** $u \leq \frac{\pi(x)}{c\tilde{\pi}(x)}$ **then**   ▷ Accept or reject proposal?
7:           $a \leftarrow$ TRUE
8:       **end if**
9:   **end while**
10:   **return** x
11: **end procedure**

$$\mathbb{P}(A = 1) = \int_\Omega \underbrace{\mathbb{P}(A = 1 \mid X = x)}_{\text{Prob line 6 of Alg 2 gives TRUE.}} \underbrace{\mathbb{P}(X = x)}_{\text{Proposal density } \tilde{\pi}.} \, dx$$

$$= \int_\Omega \underbrace{\mathbb{P}\left(U \leq \frac{\pi(x)}{c\tilde{\pi}(x)}\right)}_{\text{Uniform CDF, } \mathbb{P}(U \leq u) = F(u) = u.} \tilde{\pi}(x) \, dx$$

$$= \int_\Omega \frac{\pi(x)}{c\tilde{\pi}(x)} \tilde{\pi}(x) \, dx$$

$$= \frac{1}{c} \int_\Omega \pi(x) \, dx$$

$$= \frac{1}{c}, \tag{2.7}$$

where Ω is the support of $\tilde{\pi}(\cdot)$, $\tilde{\pi}(x) > 0$, $\forall\, x \in \Omega$. The final line follows because the integral of a density over the whole space is 1.

Hence, the number of iterations of the loop on lines 3–9 in Algorithm 2.2 which must be performed to return a single sample from $\pi(\cdot)$ is Geometrically distributed with parameter $\frac{1}{c}$. Therefore, the expected number of random number generations and function evaluations which must be performed is $2c$ per sample from $\pi(\cdot)$.

To see that Algorithm 2.2 does indeed give a sample from $\pi(\cdot)$, we note that the samples returned are only those which are accepted, so we condition on this event:

$$\mathbb{P}(X \in E \mid A = 1) = \frac{\mathbb{P}(A = 1 \mid X \in E)\,\mathbb{P}(X \in E)}{\mathbb{P}(A = 1)} \qquad \forall\, E \in \mathcal{B}$$

$$= \frac{\int_E \frac{\pi(x)}{c\tilde{\pi}(x)} \tilde{\pi}(x) \, dx}{\frac{1}{c}}$$

$$= \int_E \pi(x) \, dx.$$

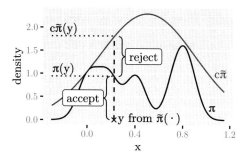

Fig. 2.3 Geometric interpretation of rejection sampling. First y is sampled from $\tilde{\pi}$ and then a uniform is sampled along the vertical dashed line at that location (i.e. between 0 and $c\tilde{\pi}(y)$). If the uniform sample falls below π then we accept and otherwise we reject. It is therefore clear that the closer $c\tilde{\pi}$ 'hugs' π the more efficient the rejection sampler

The last line is the probability of event E under the distribution with density $\pi(\cdot)$, as required.

There is a nice geometric interpretation of the rejection sampling algorithm which aids greatly with intuition. Notice that the condition in line 6 can be rewritten $uc\tilde{\pi}(x) \leq \pi(x)$. This means $uc\tilde{\pi}(x)$ is a uniform random number in the interval $[0, c\tilde{\pi}(x)]$, so that we can view rejection sampling as first drawing a value from $\tilde{\pi}(x)$, then moving it up to a uniformly distributed height under the curve $c\tilde{\pi}(x)$. The consequence of this is that we are effectively sampling uniformly points under the curve $c\tilde{\pi}(x)$ and accepting those that fall under the curve $\pi(x)$, as depicted in Fig. 2.3.

Care is required with rejection sampling in high dimensions because it is quite easy for the acceptance probability to become so small as to make the technique impractical. We will see that, as well as how to implement rejection sampling, in the following example.

2.3.2.1 Example

Consider the problem of sampling from a zero mean d-dimensional multivariate Normal distribution, having density:

$$\pi(\mathbf{x}) = (2\pi)^{-d/2} \det(\Sigma)^{-1/2} \exp\left(-\frac{1}{2}\mathbf{x}^{\mathsf{T}}\Sigma^{-1}\mathbf{x}\right), \qquad \mathbf{x} \in \mathbb{R}^d,$$

where Σ is a $d \times d$ symmetric positive semi-definite covariance matrix. It is comparatively easy to sample univariate Normal random variables (e.g. using inverse sampling in R [26] as mentioned earlier, or via a transformation type approach like [3]). Thus we could consider using a multivariate Normal with diagonal covariance,

$\sigma^2 I$, as a proposal, because this simply requires sampling d univariate Normal random variables.

This would mean we need to determine $c < \infty$ such that

$$c \det(\sigma^2 I)^{-1/2} \exp\left(-\frac{1}{2}\mathbf{x}^{\mathrm{T}}\sigma^{-2}I\mathbf{x}\right) \geq \det(\Sigma)^{-1/2} \exp\left(-\frac{1}{2}\mathbf{x}^{\mathrm{T}}\Sigma^{-1}\mathbf{x}\right) \quad \forall \mathbf{x} \in \mathbb{R}^d \tag{2.8}$$

Σ is symmetric, so it has eigendecomposition $\Sigma = Q\Lambda Q^{\mathrm{T}} \implies \Sigma^{-1} = Q\Lambda^{-1}Q^{\mathrm{T}}$, where Q is an orthogonal matrix and Λ is a diagonal matrix with entries consisting of the eigenvalues $\lambda_1, \ldots, \lambda_d$. The orthogonal transformation $\mathbf{y} = Q^{\mathrm{T}}\mathbf{x}$ also spans $\mathbb{R}^d$, so that (2.8) $\iff$

$$c\sigma^{-d} \exp\left(-\frac{1}{2}\mathbf{y}^{\mathrm{T}}\sigma^{-2}I\mathbf{y}\right) \geq \left(\prod_{i=1}^{d}\lambda_i\right)^{-1/2} \exp\left(-\frac{1}{2}\mathbf{y}^{\mathrm{T}}\Lambda^{-1}\mathbf{y}\right), \quad \forall \mathbf{y} \in \mathbb{R}^d$$

$$\iff 2\log c \geq \sum_{i=1}^{d}(\sigma^{-2} - \lambda_i^{-1})y_i^2 + 2d\log\sigma - \sum_{i=1}^{d}\log\lambda_i.$$

If $\sigma^{-2} < \lambda_i^{-1}$ for any i, then the right-hand side cannot be bounded above (since the inequality must hold $\forall \; y_i \in \mathbb{R}$), so we must have $\max_i \lambda_i < \sigma^2$ and then clearly c is minimised for $\sigma^2 = \max_i \lambda_i$. Since every term in the first sum of the right-hand side is necessarily negative, the right-hand side is maximal for $y_i = 0 \; \forall \; i$, so that the optimal c is

$$c = \left(\max_i \lambda_i\right)^{d/2} \left(\prod_{i=1}^{d}\lambda_i\right)^{-1/2},$$

when $\sigma^2 = \max_i \lambda_i$.

In summary, there is a constraint on our proposal $\tilde{\pi}(\cdot)$ when it is an uncorrelated multivariate Normal density, or else it cannot bound $\pi(\cdot)$. Moreover, we can explicitly compute the optimal proposal variance, σ^2, to give us the highest possible acceptance rate.

To make this example concrete, consider rejection sampling in this setting where

$$\Sigma = \begin{pmatrix} 1 & 0.9 & \cdots & 0.9 \\ 0.9 & 1 & \cdots & 0.9 \\ \vdots & \vdots & \ddots & \vdots \\ 0.9 & 0.9 & \cdots & 1. \end{pmatrix}$$

Note that Σ can be written as $0.1I + B$, where B is a matrix with 0.9 in every element. The rank of B is 1, so it has a single non-zero eigenvalue which must therefore equal $\mathrm{tr}(B) = 0.9d$, and the eigenvalues of $0.1I$ are all 0.1. Further-

Table 2.1 Optimal proposal variance and acceptance probability for rejection sampling a correlated multivariate Normal distribution using an uncorrelated multivariate Normal proposal

d	$\sigma^2 = \max_i \lambda_i$	Acceptance probability $\frac{1}{c}$
1	1	1
2	1.9	0.229
3	2.8	0.036
4	3.7	0.004
5	4.6	4.45×10^{-5}
$\vdots$	$\vdots$	$\vdots$
10	9.1	1.53×10^{-9}

more, $0.1I$ and B commute, therefore the eigenvalues of Σ are the sum of these eigenvalues: that is, $\lambda_1 = 0.9d + 0.1$ and $\lambda_i = 0.1 \,\forall\, i \neq 1$. As the dimension of $\pi(\cdot)$ increases, the spectral gap increases linearly and thus c grows very fast: $c = (10\lambda_1)^{(d-1)/2} = (9d + 1)^{(d-1)/2}$. Indeed, this is faster than exponential and faster than factorial growth! Consequently, for growing dimension, the acceptance probability falls super-exponentially fast—not a desirable property. See Table 2.1 for some example values.

A whimsical observation to emphasise the problem: a modern laptop can produce roughly 15 million univariate Normal samples per second and the universe is estimated to be 4.32×10^{17} seconds old. Ignoring the time to evaluate the uniform draw u or acceptance/rejection, this means the expected number of samples that would be generated by Algorithm 2.2 for this multivariate Normal problem in d-dimensions—if run for as long as the universe has existed—would be

$$\frac{1.5 \times 10^7 \times 4.32 \times 10^{17}}{d(9d + 1)^{(d-1)/2}}.$$

Consequently, even knowing the exactly optimal choice for σ^2 in our proposal, this would only be expected to render 5 samples for a 21-dimensional multivariate Normal with the innocuous looking Σ given above—rejection sampling in high dimensions can be problematic!

2.3.3 Importance Sampling

The final core standard Monte Carlo method we cover in this primer also starts from the perspective of having a proposal density $\tilde{\pi}(\cdot)$, though we no longer require it to be able to bound $\pi(\cdot)$. Importance sampling then dispenses with the notion of directly generating iid samples from $\pi(\cdot)$ and focuses on their use: in computing expectations using those samples in (2.4). Consequently, importance sampling weights the samples from $\tilde{\pi}(\cdot)$ in precisely the proportion that ensures these weighted samples produce expectations which are concordant with expectations under $\pi(\cdot)$ when used in (2.4). This is laid out precisely in Algorithm 2.3.

Algorithm 2.3 Importance sampling algorithm

1: **procedure** IMPORTANCE SAMPLING($\pi(\cdot), \tilde{\pi}(\cdot), c$)   ▷ Generate random sample from distribution with un-normalised density $\pi(\cdot)$
2:    $x \sim \tilde{\pi}(\cdot)$ ▷ Propose sample
3:    $w \leftarrow \frac{\pi(x)}{\tilde{\pi}(x)}$
4:    **return** (x, w).
5: **end procedure**

To see that this weighting has the desired effect, consider the expectation which is our objective. We first consider the situation where both π and $\tilde{\pi}$ are normalised:

$$
\begin{aligned}
\mathbb{E}_\pi[f(X)] &= \int_\Omega f(x)\pi(x)\,dx \\
&= \int_\Omega f(x)\frac{\tilde{\pi}(x)}{\tilde{\pi}(x)}\pi(x)\,dx \qquad\qquad \text{multiply and divide by } \tilde{\pi}(x) \\
&= \int_\Omega \left(f(x)\frac{\pi(x)}{\tilde{\pi}(x)} \right) \tilde{\pi}(x)\,dx \\
&= \mathbb{E}_{\tilde{\pi}} \left[\frac{f(X)\pi(X)}{\tilde{\pi}(X)} \right].
\end{aligned}
$$

That is, we use samples directly from $\tilde{\pi}(\cdot)$, and instead adjust (2.4) to target the expectation of the same functional under $\pi(\cdot)$.

$$
\mu \triangleq \int_\Omega f(x)\pi(x)\,dx \approx \frac{1}{n}\sum_{j=1}^n f(x_j)\underbrace{\frac{\pi(x_j)}{\tilde{\pi}(x_j)}}_{=w_j} = \frac{1}{n}\sum_{j=1}^n f(x_j)w_j \triangleq \hat{\mu}, \qquad (2.9)
$$

where now $x_j \sim \tilde{\pi}(\cdot)$.

Some care is required, because although this estimator remains unbiased, the variance is no longer going to be the same as the usual Monte Carlo variance where $x_j \sim \pi(\cdot)$. Indeed, now

$$
\text{Var}(\hat{\mu}) = \frac{\sigma_{\tilde{\pi}}^2}{n} \quad \text{where} \quad \sigma_{\tilde{\pi}}^2 = \int_\Omega \frac{(f(x)\pi(x) - \mu\tilde{\pi}(x))^2}{\tilde{\pi}(x)}\,dx,
$$

which can be empirically estimated from the importance samples using,

$$\hat{\sigma}_{\tilde{\pi}}^2 = \frac{1}{n} \sum_{j=1}^{n} \left(f(x_j)w_j - \hat{\mu} \right)^2. \tag{2.10}$$

As such, $\sigma_{\tilde{\pi}}^2$ (or its empirical estimate $\hat{\sigma}_{\tilde{\pi}}^2$) provide a guide to when we have a 'good' importance sampling algorithm, since with $\tilde{\pi}(\cdot)$ fixed the only option to improve the estimate is to increase the sample size n.

Indeed, it can be shown [15] that the theoretically optimal proposal distribution which minimises the estimator variance is

$$\tilde{\pi}(x)_{\text{opt}} = \frac{|f(x)|\pi(x)}{\int_{\Omega} |f(x)|\pi(x)\,dx}.$$

In particular, note that this implies that importance sampling can achieve *super-efficiency* whereby it results in lower variance even than sampling directly from $\pi(\cdot)$ when $f(x) \neq x$. Specifically, if $f(x) \geq 0 \; \forall x$ then this proposal results in a zero-variance estimator! Of course, in practice we cannot usually sample from and evaluate this optimal proposal, since it is at least as difficult as the original problem we were attempting to solve. However, even though these optimal proposals are often unusable, they provide guidance towards the form of a good proposal for any given importance sampling problem.

2.3.3.1 Self-normalising Weights

The option to use rejection sampling with un-normalised densities is very helpful (e.g. in Bayesian settings where the normalising constant is often unknown). We can retain this advantage with importance sampling by using *self-normalising weights*. The algorithm to generate the weights remains as in Algorithm 2.3, but the computation of the estimator in (2.9) changes. The self-normalised version, rather than dividing by n, uses the sum of the weights,

$$\hat{\mu}^{\star} \triangleq \frac{\sum_{j=1}^{n} f(x_j)w_j}{\sum_{j=1}^{n} w_j},$$

thereby ensuring cancellation of the unknown normalising constant from the target and/or proposal distributions in the weights.

However, it is important to note that this estimator is no longer unbiased, though asymptotically it is. Additionally, the variance of this estimator is more complicated having only approximate form. An approximate estimate can be computed using,

$$\mathrm{Var}(\hat{\mu}^\star) \approx \frac{\hat{\sigma}_{\tilde{\pi}}^{\star 2}}{n} \quad \text{where} \quad \hat{\sigma}_{\tilde{\pi}}^{\star 2} = \sum_{j=1}^{n} w_j^{\star 2} \left(f(x_j) - \hat{\mu}^\star \right)^2$$

$$\text{and} \quad w_j^\star = \frac{w_j}{\sum_{i=1}^{n} w_i}.$$

Finally, the theoretically optimal (but usually unusable) proposal in the self-normalised weight case is

$$\tilde{\pi}(x)_{\mathrm{opt}} \propto |f(x) - \mu| \pi(x).$$

In both regular and self-normalised weight settings, one can then compute appropriate confidence intervals in the usual manner.

2.3.3.2 Diagnostics

Additional care is required in the application of importance sampling when compared to using iid samples from the distribution of interest. In particular, because importance sampling uses a weighted collection of samples, it is not uncommon to be in a situation where a small number of samples with large weight dominate the estimate, so that simply having many importance samples does not equate to good estimation overall.

A common diagnostic for potential weight imbalance is derived by equating the variance of a weighted importance sampling approach to the standard iid Monte Carlo variance for an average computed using a fixed but unknown sample size n_e. Upon simple algebraic rearrangement one may then solve for n_e, the so-called *effective sample size*. This informally corresponds to the size of iid Monte Carlo sample one would expect to need to attain the same variance achieved via this importance sample, so that a low value indicates poor weight behaviour (since that corresponds to few iid samples).

$$\mathrm{Var}\left(\frac{\sum_{i=1}^{n} f(x_i) w_i}{\sum_{i=1}^{n} w_i} \right) = \frac{\sigma^2}{n_e} \quad \Longrightarrow \quad n_e = \frac{n \bar{w}^2}{\overline{w^2}},$$

where

$$\bar{w}^2 = \left(\frac{1}{n} \sum_{j=1}^{n} w_j \right)^2 \quad \text{and} \quad \overline{w^2} = \frac{1}{n} \sum_{j=1}^{n} w_j^2.$$

The reason to use such a diagnostic and not simply rely on the empirical variance estimates above is that they are themselves based on the sampling procedure and therefore may be poor estimates too.

Finally, it is critical to note that although small n_e does diagnose a problem with importance sampling, it is not necessarily true that large n_e means everything is ok:

it is, for example, entirely possible that the sampler has missed whole regions of high probability.

2.3.3.3 Example

Consider the toy problem of computing $\mathbb{P}(X > 3.09)$ when X is a univariate standard Normal random variable. The R-language [26] computes the distribution function of the standard Normal to at least 18 significant digits using a rational Chebyshev approximation [6] (see `/src/nmath/pnorm.c`) and we know the true answer to be 0.001001 (4 sf). However, if the reader suspends disbelief and imagines that we cannot accurately compute the distribution function, but can only compute the Normal density and draw random realisations from it, then evaluation of the above probability might be approximated using Monte Carlo methods instead (given the unbounded support and extreme tail location this may be preferred to numerical integration).

Since we are assuming the ability to generate random realisations from the Normal distribution, a standard Monte Carlo approach would draw many samples $x_i \sim \mathrm{N}(0, 1)$ and compute

$$\mathbb{P}(X > 3.09) = \mathbb{E}\left[\mathbb{I}_{[3.09,\infty)}(X)\right] = \frac{1}{n} \sum_{j=1}^{n} \mathbb{I}_{[3.09,\infty)}(x_j)$$

However, this will require many samples to achieve an accurate estimate of this tail probability.

In contrast, still only using simulations from a Normal distribution, we may elect to use importance sampling with a proposal $\mathrm{N}(m, 1)$ for some choice m. We know the fully normalised density of a Normal distribution and therefore will be using the estimator (2.9) with associated single sample variance which can be approximated using (2.10). Therefore, to select m, we perform a small grid search over possible proposals, computing $\hat{\sigma}_{\tilde{\pi}}^2$ each time, to find a good choice. This results in Fig. 2.4, showing that a proposal $\mathrm{N}(3.25, 1)$ is a good choice.

A further final run of $n = 100{,}000$ samples renders an estimate $\hat{\mu} = 0.001002$ (4 sf). The same pseudo-random number stream using standard Monte Carlo renders an estimate 0.001140 (4 sf), which is a relative error $163\times$ larger than the importance sampling estimate. To demonstrate this is not a 'fluke' result, we continue to repeat both importance sampling and standard Monte Carlo estimation with runs of size $n = 100{,}000$ and plot the density of estimates of $\mathbb{P}(X > 3.09)$ in Fig. 2.5.

Note that Fig. 2.5 demonstrates how much more accurate importance sampling is for the same sample size $n = 100{,}000$ when computing this event probability compared to standard Monte Carlo. One may reasonably object that we have ignored the 25 pilot runs of $n = 100{,}000$ importance samples used to select $m = 3.25$, so that the total computational effort expended on importance sampling was at least $26\times$ that of standard Monte Carlo. However, it is a simple calculation to determine

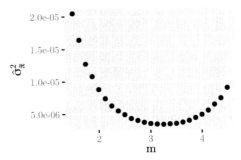

Fig. 2.4 The estimate of $\hat{\sigma}_{\hat{\pi}}^2$ using (2.10) for 25 different values of m in the Normal proposal $N(m, 1)$ used in an importance sampling estimator of $\mathbb{P}(X > 3.09)$, where $X \sim N(0, 1)$. Each estimate of $\hat{\sigma}_{\hat{\pi}}^2$ is based on $n = 100,000$ samples. m varies on an equally spaced grid from 1.5 to 4.5. The minimum is at $m = 3.25$

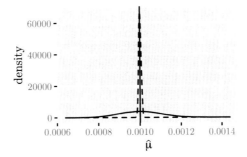

Fig. 2.5 A total of $10,000$ runs of both importance sampling and standard Monte Carlo, each of size $n = 100,000$. Each run was used to compute the estimate of $\mathbb{P}(X > 3.09)$ where $X \sim N(0, 1)$ and a kernel density plot of these estimates produced (importance sampling = dashed, standard Monte Carlo = solid). The vertical line is the ground truth computed using `pnorm(3.09, lower.tail = FALSE)`. The same pseudo-random number stream was used for each method to ensure a fair comparison

that based on the standard deviation of the samples used to generate Fig. 2.5 and the $\sqrt{n}$ convergence of Monte Carlo, that it would require a standard Monte Carlo sample of size $n = 295 \times 100,000$ to achieve the same accuracy profile as importance sampling. Therefore, even accounting for the pilot computational effort to select a proposal distribution, there is a substantial benefit to using importance sampling.

2.4   Further Reading

A textbook length introduction with a solid emphasis on implementation details in R can be found in [28]. The same authors have a more advanced textbook going into the theoretical aspects more deeply [27]. Both these books also introduce Markov

Chain Monte Carlo methods, which are often used in practice in high dimensional problems. A nice tutorial paper introduction to MCMC is [1] and [4] is an excellent collection of chapters on the topic.

A classic Monte Carlo text is [10], which is now freely (and legally) available online and contains many results not easily found elsewhere.

Although standard Monte Carlo and Markov Chain Monte Carlo arguably represent the mainstay of most practical uses of Monte Carlo, there are an array of advanced methods which are particularly well suited to different settings. Some excellent review texts as jumping off points to explore some of these include [8] (Sequential Monte Carlo), [19, 20] (Approximate Bayesian Computation), [14] (Multi-Level Monte Carlo), and [2] (MLMC in engineering reliability).

Excellent software choices for practically performing inference in complex models via sampling methods includes Stan [5] and Birch [23].

Dedication

This chapter is dedicated to the memory of Brett Houlding (1982–2019). The tragic news of Brett's passing was received while I was in the act of writing this chapter. He will be sorely missed.

References

1. C. Andrieu, N. de Freitas, A. Doucet, and M. I. Jordan. An introduction to MCMC for Machine Learning. *Machine Learning*, 50:5–43, 2003.
2. L. J. M. Aslett, T. Nagapetyan, and S. J. Vollmer. Multilevel Monte Carlo for Reliability Theory. *Reliability Engineering & System Safety*, 165:188–196, 2017.
3. G. E. P. Box and M. E. Muller. A note on the generation of Normal deviates. *Annals of Mathematical Statistics*, 29(2):610–611, 1958.
4. S. Brooks, A. Gelman, G. Jones, and X.-L. Meng, editors. *Handbook of Markov Chain Monte Carlo*. Chapman and Hall/CRC, 2011.
5. B. Carpenter, A. Gelman, M. D. Hoffman, D. Lee, B. Goodrich, M. Betancourt, M. Brubaker, J. Guo, P. Li, and A. Riddell. Stan: A probabilistic programming language. *Journal of Statistical Software*, 76(1):1–32, 2017. https://mc-stan.org/.
6. W. J. Cody. Rational Chebyshev approximations for the error function. *Mathematics of Computation*, 23(107):631–637, 1969.
7. D. R. Cox. Applied statistics: a review. *The Annals of Applied Statistics*, 1(1):1–16, 2007.
8. C. Dai, J. Heng, P. E. Jacob, and N. Whiteley. An invitation to sequential Monte Carlo samplers, 2020. arXiv:2007.11936 [stat.CO].
9. H. Dai, M. Pollock, and G. O. Roberts. Monte Carlo Fusion. *arXiv preprint* arXiv:1901.00139, 2019.
10. L. Devroye. *Non-Uniform Random Variate Generation*. Springer-Verlag, New York, 1986. Available at http://luc.devroye.org/rnbookindex.html.
11. G. Feng, E. Patelli, M. Beer, and F. P. A. Coolen. Imprecise system reliability and component importance based on survival signature. *Reliability Engineering & System Safety*, 150:116–125, 2016.

12. N. Friel, R. Rastelli, J. Wyse, and A. E. Raftery. Interlocking directorates in Irish companies using a latent space model for bipartite networks. *Proceedings of the National Academy of Sciences*, 113(24):6629–6634, 2016.
13. J. E. Gentle. *Random number generation and Monte Carlo methods*. Springer Science & Business Media, 2006.
14. M. B. Giles. Multilevel Monte Carlo methods. *Acta Numerica*, 24:259–328, 2015.
15. G. Goertzel. *Quota sampling and importance functions in stochastic solution of particle problems*, volume 2793. US Atomic Energy Commission, Technical Information Division, 1950.
16. J. Haslett, M. Whiley, S. Bhattacharya, M. Salter-Townshend, S. P. Wilson, J. R. M. Allen, B. Huntley, and F. J. G. Mitchell. Bayesian palaeoclimate reconstruction. *Journal of the Royal Statistical Society: Series A (Statistics in Society)*, 169(3):395–438, 2006.
17. B. Houlding and S. P. Wilson. Considerations on the UK re-arrest hazard data analysis. *Law, Probability & Risk*, 10(4):303–327, 2011.
18. A. J. Kinderman and J. G. Ramage. Computer generation of Normal random variables. *Journal of the American Statistical Association*, 71(356):893–896, 1976.
19. J. M. Marin, P. Pudlo, C. P. Robert, and R. J. Ryder. Approximate Bayesian computational methods. *Statistics and Computing*, 22(6):1167–1180, 2012.
20. P. Marjoram, J. Molitor, V. Plagnol, and S. Tavaré. Markov chain Monte Carlo without likelihoods. *PNAS*, 100(26):15324–15328, 2003.
21. G. Marsaglia. Random number generators. *Journal of Modern Applied Statistical Methods*, 2(1):2, 2003.
22. M. Matsumoto and T. Nishimura. Mersenne twister: a 623-dimensionally equidistributed uniform pseudo-random number generator. *ACM Transactions on Modeling and Computer Simulation (TOMACS)*, 8(1):3–30, 1998.
23. L. M. Murray and T. B. Schön. Automated learning with a probabilistic programming language: Birch. *Annual Reviews in Control*, 46:29–43, 2018. https://www.birch.sh/.
24. A. B. Owen. *Monte Carlo theory, methods and examples*. 2013.
25. M. Pollock, A. M. Johansen, and G. O. Roberts. On the exact and ε-strong simulation of (jump) diffusions. *Bernoulli*, 22(2):794–856, 2016.
26. R Core Team. *R: A Language and Environment for Statistical Computing*. R Foundation for Statistical Computing, Vienna, Austria, 2019.
27. C. P. Robert and G. Casella. *Monte Carlo Statistical Methods*. Springer-Verlag New York, 2004.
28. C. P. Robert and G. Casella. *Introducing Monte Carlo Methods with R*. Springer, New York, NY, 2010.
29. S. P. Wilson and M. J. Costello. Predicting future discoveries of European marine species by using a non-homogeneous renewal process. *Journal of the Royal Statistical Society: Series C*, 54(5):897–918, 2005.
30. E. Zio, P. Baraldi, and E. Patelli. Assessment of the availability of an offshore installation by Monte Carlo simulation. *International Journal of Pressure Vessels and Piping*, 83(4):312–320, 2006.

Chapter 3
Introduction to the Theory of Imprecise Probability

Erik Quaeghebeur

Abstract The theory of imprecise probability is a generalization of classical 'precise' probability theory that allows modeling imprecision and indecision. This is a practical advantage in situations where a unique precise uncertainty model cannot be justified. This arises, for example, when there is a relatively small amount of data available to learn the uncertainty model or when the model's structure cannot be defined uniquely. The tools the theory provides make it possible to draw conclusions and make decisions that correctly reflect the limited information or knowledge available for the uncertainty modeling task. This extra expressivity however often implies a higher computational burden. The goal of this chapter is to primarily give you the necessary knowledge to be able to read literature that makes use of the theory of imprecise probability. A secondary goal is to provide the insight needed to use imprecise probabilities in your own research. To achieve the goals, we present the essential concepts and techniques from the theory, as well as give a less in-depth overview of the various specific uncertainty models used. Throughout, examples are used to make things concrete. We build on the assumed basic knowledge of classical probability theory.

3.1 Introduction

The theory of imprecise probability is a generalization of classical 'precise' probability theory that allows modeling imprecision and indecision. Why is such a theory necessary? Because in many practical applications a lack of information—e.g., about model parameters—and paucity of data—especially if we also consider conditional models—make it impossible to create a reliable model.

For example, consider a Bayesian context where a so-called prior probability distribution must be chosen as part of the modeling effort. The lack of information

---

(This chapter was written while at Delft University of Technology)

E. Quaeghebeur (✉)
Eindhoven University of Technology, Eindhoven, The Netherlands

© The Author(s) 2022
L. Aslett et al. (eds.), *Uncertainty in Engineering*,
SpringerBriefs in Statistics,
https://doi.org/10.1007/978-3-030-83640-5_3

may make it difficult to determine the type of the prior distribution, let alone its parameters. Then, even if we assume some prior has been chosen—e.g., a normal one—in a somewhat arbitrary way, a paucity of data will make the parameters of the posterior—updated—distribution depend to a large degree on the prior's somewhat arbitrary parameters. The consequence is that conclusions drawn from the posterior are unreliable and decisions based on it somewhat arbitrary.

The theory of imprecise probability provides us with a set of tools for dealing with the problem described above. For the example above, instead of choosing a single prior distribution, a whole set of priors is used, one that is large enough to sufficiently reduce or even eliminate the arbitrariness of this modeling step. The consequence is that conclusions drawn from an imprecise probabilistic model are more reliable by being less committal—more vague, if you wish; some would say 'more honest'—and that decisions based on it allow for indecision.

In this chapter, we will go over the basic concepts of the theory of imprecise probability theory. Therefore, we will consider 'small' problems, with finite possibility spaces. However, the theory can be applied to infinite—countable and uncountable—possibility spaces as well. Also, only the basics of more advanced topics such as conditioning will be touched upon. But, and this is the chapter's goal, after having understood the material we do treat, the imprecise probability literature should have become substantially more accessible. Good extensive general treatments are available [2, 16, 20] and the proceedings of the ISIPTA conferences provide an extensive selection of papers developing imprecise probability theory or applying it [1, 3, 4, 6–12].

Concretely, we start with a discussion of the fundamental concepts in Sect. 3.2. This is done in terms of the more basic notion of sets of acceptable gambles. Probabilities only appear thereafter, in Sect. 3.3, together with the related notion of prevision (expectation). The connection with sets of probabilities is made next, in Sect. 3.4. Then we touch upon conditioning, in Sect. 3.5, and before closing add some remarks about continuous possibility spaces, in Sect. 3.6. Throughout we will spend ample time on a running example to illustrate the theory that is introduced.

3.2 Fundamental Concepts

In this section, we introduce the fundamental concepts of the theory of imprecise probability [18] [20, §3.7]. First, in Sect. 3.2.1, we get started with some basic concepts. Then, in Sect. 3.2.2, we list and discuss the coherence criteria on which the whole theory is built.

3.2.1 Basic Concepts

Consider an *agent* reasoning about an *experiment* with an uncertain *outcome*. This experiment is modeled using a *possibility space*—a set—$\mathcal{X}$ of outcomes x. Now consider the linear space $\mathcal{L} = \mathcal{X} \to \mathbb{R}$ of real-valued functions over the outcomes. We view these functions as *gambles* because they give a value, seen as a *payoff*, for each outcome and because the outcome is uncertain and therefore the payoff is as well. A special class of gambles are the outcome *indicators* 1_x or subset *indicators* 1_B, which take the value one on that outcome or subset and zero elsewhere.

The agent can then express her uncertainty by specifying a set of gambles, called an *assessment* $\mathcal{A}$, that she considers *acceptable*. Starting from such an assessment, she can reason about other gambles and decide whether she should also accept them or not. If she were to do this for all gambles, then the *natural extension* $\mathcal{E}$ of her assessment would be the set of all acceptable gambles. To reason in a principled way, she needs some guiding criteria; these are the next section's topic.

Let us now introduce our running example:

Wiske and Yoko Tsuno want to bet on Belgium vs. Japan

Given a sports match between *Belgium* and *Japan*, there is uncertainty about which country's team will win. So we consider the possibility space $\{\text{BE}, \text{JP}\}$. There are to agents—gamblers—: *Wiske* and *Yoko Tsuno*, two comic book heroines. Each has an assessment consisting of a single gamble that they find acceptable:

- Wiske accepts losing 5 coins if Japan wins for the opportunity to win 1 coin if Belgium wins; so $\mathcal{A}_W = \{1_{\text{BE}} - 5 \cdot 1_{\text{JP}}\}$.
- Yoko Tsuno accepts losing 4 coins if belgium wins for the opportunity to win 1 coin if Japan wins; so $\mathcal{A}_Y = \{-4 \cdot 1_{\text{BE}} + 1_{\text{JP}}\}$.

The heroines are also discussing joining forces and forming a betting pool. The pools they consider are

- 'Simple', formed by combining their assessments; so

$$\mathcal{A}_{SP} = \{1_{\text{BE}} - 5 \cdot 1_{\text{JP}}, -4 \cdot 1_{\text{BE}} + 1_{\text{JP}}\}.$$

- 'Empty' in case of disagreement, without any acceptable gambles; so $\mathcal{A}_{EP} = \emptyset$.

3.2.2 Coherence

In the theory of imprecise probabilities, the classical rationality criteria used for reasoning about assessments are called *coherence* criteria. These are typically formulated as four rules that should apply to any gambles f and g. (There are different

variants in the literature, but the differences are not relevant in this introductory text.)
We divide the criteria into two classes.

Constructive   State how to generate acceptable gambles from the assessment:
  Positive scaling If f is acceptable and $\lambda > 0$, then $\lambda \cdot f$ is acceptable.
      Addition If f and g are acceptable, then $f + g$ is acceptable.

Background   State which gambles are always or never acceptable:
    Accepting gain If f is nonnegative for all outcomes, then f is acceptable.
  Avoiding sure loss If g is negative for all outcomes, then g is not acceptable.

These criteria are quite broadly seen as reasonable, under the assumption that the
payoffs are 'not too large'.

The last criterion, 'Avoiding sure loss', puts a constraint on what is considered
coherent; if it is violated, we say that an assessment *incurs sure loss*. The first three
rules can be used to create an explicit expression for the natural extension:

$$\mathcal{E} = \left\{ \textstyle\sum_{f \in \mathcal{K}} \lambda_f \cdot f : \mathcal{K} \Subset \mathcal{A} \cup \{ f \in \mathcal{L} : f \geq 0 \} \text{ and } (\forall f \in \mathcal{K} : \lambda_f \geq 0) \right\},$$

where $\Subset$ denotes the finite subset relation. Then $\mathcal{E}$ is the smallest convex cone of
gambles encompassing the assessment $\mathcal{A}$ and the nonnegative gambles—including
the zero gamble.

Let us apply the natural extension to our running example:

The natural extensions of Wiske, Yoko Tsuno, and the betting pools

For our finite possibility space,

$$\{ f \in \mathcal{L} : f \geq 0 \} = \left\{ \textstyle\sum_{x \in \mathcal{X}} \mu_x \cdot 1_x : (\forall x \in \mathcal{X} : \mu_x \geq 0) \right\}$$

So, with $\lambda_A, \mu_x \geq 0$ for all outcomes x and agent identifiers A, we get the following
expressions that characterize the natural extensions:

$$\text{Wiske} \quad \begin{aligned} \lambda_W \cdot (1_{BE} - 5 \cdot 1_{JP}) &+ \mu_{BE} \cdot 1_{BE} + \mu_{JP} \cdot 1_{JP} \\ &= (\lambda_W + \mu_{BE}) \cdot 1_{BE} + (-5 \cdot \lambda_W + \mu_{JP}) \cdot 1_{JP}, \end{aligned}$$

$$\text{Yoko Tsuno} \quad \begin{aligned} \lambda_Y \cdot (-4 \cdot 1_{BE} + 1_{JP}) &+ \mu_{BE} \cdot 1_{BE} + \mu_{JP} \cdot 1_{JP} \\ &= (-4 \cdot \lambda_Y + \mu_{BE}) \cdot 1_{BE} + (\lambda_Y + \mu_{JP}) \cdot 1_{JP}, \end{aligned}$$

Simple pool $(\lambda_W - 4 \cdot \lambda_Y + \mu_{BE}) \cdot 1_{BE} + (-5 \cdot \lambda_W + \lambda_Y + \mu_{JP}) \cdot 1_{JP}$,

Empty pool $\mu_{BE} \cdot 1_{BE} + \mu_{JP} \cdot 1_{JP}$.

To check whether the natural extension incurs sure loss, we must check whether the
coefficients of 1_{BE} and 1_{JP} can become negative at the same time. Only the simple pool
incurs sure loss; e.g., fill in $\lambda_W = \lambda_Y = 1$ and $\mu_{BE} = \mu_{JP} = 0$ to convince yourself.
(Convince yourself as well that the others avoid sure loss indeed.)

3.3 Previsions and Probabilities

In this section, we move from modeling uncertainty using sets of acceptable gambles to the more familiar language of expectation—or, synonymously, *prevision*—
and probability [17]. We first transition from acceptable gambles to previsions in
Sect. 3.3.1 [18, §1.6.3] [17, §2.2] and in a second step, in Sect. 3.3.2, give the connection to probabilities [20, §2.6]. Next, in Sect. 3.3.3, we consider assessments in
terms of previsions and what the other fundamental concepts of Sect. 3.2 then look
like [17, §2.2.1, §2.2.4] [20, §2.4–5, §3.1]. Finally, in Sect. 3.3.4, we consider the
important special case of assessments in terms of previsions defined on a linear space
of gambles [17, §2.2.1] [20, §2.3.2–6].

3.3.1 Previsions as Prices for Gambles

Before we start: 'prevision' is in much of the imprecise probability literature used
as a synonym for 'expectation'; we here follow that tradition.

  Now, how do we get an agent's *previsions* for a gamble—equivalently: expectation of a random variable—given that we know the agent's assessment as a set of
acceptable gambles $\mathcal{A}$? We first define a *price* to be a constant gamble and identify this constant gamble with its constant payoff value. Then we define the agent's
previsions as specific types of *acceptable* prices:

- The *lower prevision* $\underline{P}(f)$ is the supremum acceptable buying price of f:

$$\underline{P}(f) = \sup\{v \in \mathbb{R} : f - v \in \mathcal{E}\}.$$

- The *upper prevision* $\overline{P}(f)$ is the infimum acceptable selling price of f:

$$\overline{P}(f) = \inf\{\kappa \in \mathbb{R} : \kappa - f \in \mathcal{E}\}.$$

If $\mathcal{E}$ is coherent, then $\underline{P}$ and $\overline{P}$ are also called coherent. There is a *conjugacy* relation
between coherent lower and upper previsions: $\overline{P}(f) = -\underline{P}(-f)$. It allows us to
work in terms of either type of prevision; we will mainly use the lower one.

  In case $\underline{P}(f) = \overline{P}(f)$, then $P(f) = \underline{P}(f)$ is the called the (precise) *prevision* of
the gamble f.

3.3.2 Probabilities as Previsions of Indicator Gambles

Now that we have definitions for lower and upper previsions, we can derive probabilities from those. For classical probability, we have that the probability of an event—a

subset B of the possibility space $\mathcal{X}$—is the prevision of the indicator for that event. For lower and upper previsions, we get:

- The *lower probability*: $\underline{P}(B) = \underline{P}(1_B)$.
- The *upper probability*: $\overline{P}(B) = \overline{P}(1_B)$.

Notice that we reuse the same symbol for the prevision and probability functions, as is common in the literature. As long as the nature of the argument—gamble or event—is clear, this does not cause ambiguity. If $\underline{P}$ and $\overline{P}$ are coherent as previsions, then so are they as probabilities. Also the conjugacy relationship can be translated to coherent lower and upper probabilities; let $B^c = \mathcal{X} \setminus B$, then

$$\overline{P}(B) = \overline{P}(1_B) = \overline{P}(1 - 1_{B^c}) = -\underline{P}(-1 + 1_{B^c}) = 1 - \underline{P}(1_{B^c}) = 1 - \underline{P}(B^c).$$

In case $\underline{P}(B) = \overline{P}(B)$, then $P(B) = \underline{P}(B)$ is called the (precise) *probability* of B.

To make the definitions for lower and upper previsions and probabilities concrete, let us apply them to our running example:

Lower and upper probabilities for all events and agents

We work out the calculation of Wiske's lower probability that Belgium will win.

$$
\begin{aligned}
\underline{P}_W(\mathrm{BE}) &= \underline{P}_W(1_{\mathrm{BE}}) \quad \text{(def. lower probability)}\\
&= \sup\left\{ v \in \mathbb{R} : 1_{\mathrm{BE}} - v \in \mathcal{E}_W \right\} \quad \text{(def. lower prevision)}\\
&= \sup\left\{ v \in \mathbb{R} : \begin{bmatrix} 1-v \\ 0-v \end{bmatrix} = \begin{bmatrix} \lambda_W + \mu_{\mathrm{BE}} \\ -5 \cdot \lambda_W + \mu_{\mathrm{JP}} \end{bmatrix}, \lambda_W \geq 0, \mu_{\mathrm{BE}} \geq 0, \mu_{\mathrm{JP}} \geq 0 \right\}\\
&\qquad\qquad \text{(write out natural extension } \mathcal{E}_W \text{ of } \mathcal{A}_W)\\
&= \sup\left\{ 5 \cdot \lambda_W - \mu_{\mathrm{JP}} : 1 - 5 \cdot \lambda_W + \mu_{\mathrm{JP}} = \lambda_W + \mu_{\mathrm{BE}}, \lambda_W \geq 0, \mu_{\mathrm{BE}} \geq 0, \mu_{\mathrm{JP}} \geq 0 \right\}\\
&\qquad\qquad \text{(eliminate } v)\\
&= \sup\left\{ 5 \cdot \lambda_W - \mu_{\mathrm{JP}} : \lambda_W = \tfrac{1}{6}(1 + \mu_{\mathrm{JP}} - \mu_{\mathrm{BE}}), \lambda_W \geq 0, \mu_{\mathrm{BE}} \geq 0, \mu_{\mathrm{JP}} \geq 0 \right\}\\
&\qquad\qquad \text{(solve constraint for } \lambda_W)\\
&= \sup\left\{ \tfrac{5}{6} - \tfrac{1}{6}\mu_{\mathrm{JP}} - \tfrac{5}{6}\mu_{\mathrm{BE}} : 1 + \mu_{\mathrm{JP}} \geq \mu_{\mathrm{BE}}, \mu_{\mathrm{BE}} \geq 0, \mu_{\mathrm{JP}} \geq 0 \right\} \quad \text{(eliminate } \lambda_W)\\
&= \frac{5}{6} \quad \text{(feasible solution } \mu_{\mathrm{BE}} = 0, \mu_{\mathrm{JP}} = 0 \text{ maximizes expression)}
\end{aligned}
$$

Do the calculations also for the other agents and Japan. Then apply conjugacy to find the following table of lower and upper probabilities:

Agents	$\underline{P}(\mathrm{BE})$, $\overline{P}(\mathrm{BE})$	$\underline{P}(\mathrm{JP})$, $\overline{P}(\mathrm{JP})$	Note
Wiske	5/6 , 1	0 , 1/6	Will not bet against Belgium
Yoko Tsuno	0 , 1/5	4/5 , 1	Will not bet against Japan
Simple pool	$+\infty$, $-\infty$	$+\infty$, $-\infty$	Sure loss, so absurd bounds
Empty pool	0 , 1	0 , 1	So-called *vacuous* model

While in the above example the calculation of the lower prevision can be done by hand, in general it realistically requires a linear program solver.

3.3.3 Assessments of Lower Previsions

Up until now, we assumed a set of acceptable gambles $\mathcal{A}$—an agent's assessment—to be given. But often the agent will directly specify lower and upper probabilities or previsions, e.g., as bounds on precise probabilities and previsions. However, the coherence criteria and expression for the natural extension are based on having a set of acceptable gambles. In this section we will provide expressions based on an assessment specified as lower prevision values for gambles in a given set $\mathcal{K}$.

The approach is to derive an assessment as a set of acceptable gambles $\mathcal{A}$ from these lower prevision. Irrespective of what its natural extension $\mathcal{E}$ actually looks like, it follows from the definition of the lower prevision as a supremum acceptable buying price that

$$0 \leq \sup\left\{v - \underline{P}(f) : v \in \mathbb{R} \wedge f - v \in \mathcal{E} \supseteq \mathcal{A}\right\}$$
$$= \sup\left\{\kappa \in \mathbb{R} : f - (\kappa + \underline{P}(f)) \in \mathcal{E}\right\} = \sup\left\{\kappa \in \mathbb{R} : (f - \underline{P}(f)) - \kappa \in \mathcal{E}\right\}.$$

This implies that $f - \underline{P}(f) + \varepsilon \in \mathcal{E}$ for any $\varepsilon > 0$, because of coherence. We cannot take $\varepsilon = 0$, because the corresponding so-called *marginal gamble* $f - \underline{P}(f)$ is not included in $\mathcal{E}$ in general, as the supremum value $\kappa = 0$ is not necessarily attained inside the set. We therefore take $\mathcal{A} = \bigcup_{f \in \mathcal{K}}\left\{f - \underline{P}(f) + \varepsilon : \varepsilon > 0\right\}$.

We can then apply the theory described above to this assessment $\mathcal{A}$. This leads to the following nontrivial results for a lower prevision $\underline{P}$ defined on a set of gambles $\mathcal{K}$:

- It *avoids sure loss* if and only if for all $n \geq 0$ and $f_k \in \mathcal{K}$ it holds that

$$\sup_{x \in X} \sum_{k=1}^{n}\left(f_k(x) - \underline{P}(f_k)\right) \geq 0.$$

- It is *coherent* if and only if for all $n, m \geq 0$ and $f_k \in \mathcal{K}$ it holds that

$$\sup_{x \in X}\left(\sum_{k=1}^{n}\left(f_k(x) - \underline{P}(f_k)\right) - m \cdot (f_0 - \underline{P}(f_0))\right) \geq 0.$$

- Its *natural extension* to any gamble f in $\mathcal{L}$ is

$$\underline{E}(f) = \sup\left\{\inf_{x \in X}\left\{f(x) - \sum_{k=1}^{n}\lambda_k \cdot \left(f_k(x) - \underline{P}(f_k)\right)\right\} : n \geq 0, f_k \in \mathcal{K}, \lambda_k \geq 0\right\}.$$

3.3.4   Working on Linear Spaces of Gambles

The coherence criterion for lower previsions on an arbitrary space $\mathcal{K}$ of gambles we gave in the preceding section is quite involved. However, in case $\mathcal{K}$ is a *linear space* of gambles, this criterion becomes considerably simpler. Namely, a lower prevision $\underline{P}$ must then satisfy the following criteria for all gambles f and g in $\mathcal{K}$ and $\lambda > 0$:

$$\begin{aligned}
\textit{Accepting sure gains} \quad & \underline{P}(f) \geq \inf f, \\
\textit{Super-linearity} \quad & \underline{P}(f + g) \geq \underline{P}(f) + \underline{P}(g), \\
\textit{Positive homogeneity} \quad & \underline{P}(\lambda f) = \lambda \cdot \underline{P}(f).
\end{aligned}$$

Expressed for upper previsions $\overline{P}$, these coherence criteria are very similar:

$$\begin{aligned}
\textit{Accepting sure gains} \quad & \overline{P}(f) \leq \sup f, \\
\textit{Sub-linearity} \quad & \overline{P}(f + g) \leq \overline{P}(f) + \overline{P}(g), \\
\textit{Positive homogeneity} \quad & \overline{P}(\lambda f) = \lambda \cdot \overline{P}(f).
\end{aligned}$$

From the coherence criteria, many useful properties can be derived for a coherent lower prevision $\underline{P}$ and its conjugate upper prevision $\overline{P}$. We provide a number of key ones, which hold for all gambles f and g in $\mathcal{K}$ and $\mu \in \mathbb{R}$; $\underline{\overline{P}}$ denotes either $\underline{P}$ or $\overline{P}$:

$$\begin{aligned}
\textit{Upper dominates lower} \quad & \overline{P}(f) \geq \underline{P}(f), \\
\textit{Constants} \quad & \underline{\overline{P}}(\mu) = \mu, \\
\textit{Constant additivity} \quad & \underline{\overline{P}}(f + \mu) = \underline{\overline{P}}(f) + \mu, \\
\textit{Gamble dominance} \quad & \text{if } f \geq g + \mu \text{ then } \underline{\overline{P}}(f) \geq \underline{\overline{P}}(g) + \mu, \\
\textit{Mixed sub/super-additivity} \quad & \underline{P}(f + g) \leq \underline{P}(f) + \overline{P}(g) \leq \overline{P}(f + g).
\end{aligned}$$

3.4   Sets of Probabilities

In Sect. 3.2 we modeled uncertainty using a set of acceptable gambles. In Sect. 3.3 we showed how this can also be done in terms of lower or upper previsions (or probabilities). In this section, we add a third representation, one using *credal sets*— sets of precise previsions [17, §2.2.2], [18, §1.6.2]. In Sect. 3.4.1 we show how to derive the credal set corresponding to a given lower prevision. In Sect. 3.4.2 we go the other direction and show how to go from a credal set to lower prevision values [20, §3.3].

3.4.1   From Lower Previsions to Credal Sets

A credal set is a subset of the set of all precise previsions $\mathcal{P}$. (For possibility spaces B different from X, we write $\mathcal{P}_B$.) This set is convex, meaning that any convex mixture of precise previsions is again a precise prevision. Because of this, a gamble's prevision is a linear function over this space. A *lower*—and *upper*—prevision can be seen as providing a bound on the value of the precise prevision for that gamble and thereby represent a *linear constraint* on the precise previsions. So the credal set $\mathcal{M}$ corresponding to a lower prevision $\underline{P}$ defined on a set of gambles $\mathcal{K}$ is the subset of $\mathcal{P}$ satisfying this constraint for all gambles in $\mathcal{K}$:

$$\mathcal{M} = \bigcap_{f \in \mathcal{K}} \{P \in \mathcal{P} : P(f) \geq \underline{P}(f)\}.$$

Being defined as such an intersection, *such* credal sets are closed and convex.

The rationality criteria for a lower prevision $\underline{P}$ we encountered before can also be expressed using its corresponding credal set $\mathcal{M}$:

- $\underline{P}$ incurs sure loss if and only if $\mathcal{M}$ is equal to the empty set.
- $\underline{P}$ is coherent if and only if all constraints are 'tight', i.e., if there exists a P in $\mathcal{M}$ such that $P(f) = \underline{P}(f)$ for all f in $\mathcal{K}$.

Let us make the concept of a credal set concrete using our running example:

Yoko Tsuno's credal set

For a finite possibility space such as the one of our running example, a precise prevision P can be defined completely by the corresponding *probability mass function p* defined by $p_x = P(\{x\})$ for x in $X = \{\text{BE, JP}\}$. The set of all precise previsions can therefore be represented by the *probability simplex*—the set of all probability mass functions—on X. This set and the example probability mass function $(\frac{1}{2}, \frac{1}{2})$ is shown below left. Below right, we illustrate how Yoko Tsuno's lower prevision $\underline{P}_Y(\text{JP}) = \frac{4}{5}$ generates the credal set $\mathcal{M}_Y$: The gamble 1_{JP} as a linear function over the simplex is shown as an inclined line. This linear relationship between p—equivalently, the corresponding prevision P_p—and $P_p(1_{\text{JP}}) = P_p(\text{JP})$ transforms the bounds $\frac{4}{5} \leq P_p(\text{JP}) \leq 1$ into $\mathcal{M}_Y$.

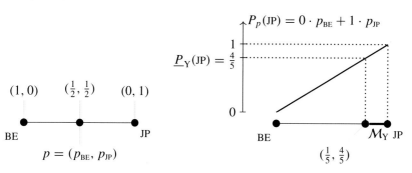

The set of extreme points $\mathcal{M}_Y^*$ of $\mathcal{M}_Y$ as probability mass functions is $\{(\frac{1}{5}, \frac{4}{5}), (0, 1)\}$.

3.4.2 From Credal Sets to Lower Previsions

Now we assume that the agent's credal set $\mathcal{M}$ is given. Most generally this can be any set of precise previsions, i.e., any subset of $\mathcal{P}$. Often, to ensure equivalence between *coherent* lower previsions and *non-empty* credal sets, they are required to be closed and convex. In that case, a credal set is determined completely by its set of *extreme points* $\mathcal{M}^*$ in the sense that all other elements are convex mixtures of these.

To determine the lower prevision corresponding to any credal set, we determine its value for each gamble f of interest using the *lower envelope theorem*:

$$\underline{P}(f) = \min \left\{ P_p(f) : p \in \mathcal{M} \right\} = \min \left\{ P_p(f) : p \in \mathcal{M}^* \right\}.$$

Let us again use the running example to provide a feeling for what this all means:

A credal set for the empty pool facing penalties

Consider the empty pool. Because its assessment is empty, its credal set $\mathcal{M}_{EP}$ is the trivial one corresponding to all probability mass functions on $\mathcal{X} = \{BE, JP\}$. Now we add an extra element to the possibility space, 'Penalties'. Below left we show $\mathcal{M}_{EP}$ embedded in the corresponding larger probability simplex. Wiske and Yoko Tsuno decide to add the uniform probability mass function to it. Below right, you see the convex hull $\mathcal{M}_{EUP}$ of this extra probability mass function and the original credal set.

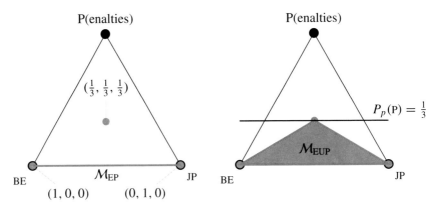

If we want to calculate lower and upper prevision values, we can here use the extreme point version of the lower and—similar—upper envelope theorem. For example, for the pool's upper probability for Penalties:

$$\overline{P}_{EP}(P) = \overline{P}_{EP}(1_P) = \max \left\{ p^\top(0,0,1) : p \in \left\{ (1,0,0), (0,1,0), (\tfrac{1}{3}, \tfrac{1}{3}, \tfrac{1}{3}) \right\} \right\} = \frac{1}{3}.$$

To make it explicit where this maximum is achieved, we above right show the line of probability mass functions p such that $P_p(P) = \frac{1}{3}$.

3.5 Basics of Conditioning

Conditioning an uncertainty model is the act of restricting attention to a subset B of the possibility space. It is often used to *update* an uncertainty model after having observed the event B [20, §6.1].

In the theory of imprecise probability, conditioning is a specific case of natural extension [17, §2.3.3], [20, §6.4.1]. In terms of acceptable gambles, conditioning on B corresponds to restricting the space of gambles to those that are zero outside B [18, §1.3.3]. For lower previsions, this translates to the following conditioning rule for all gambles f in $\mathcal{L}$:

$$\underline{E}(f \mid B) = \begin{cases} \inf_{x \in B} f(x) & \text{if } \underline{P}(B) = 0, \\ \max\{\mu \in \mathbb{R} : \underline{P}(1_B(f - \mu)) = 0\} & \text{if } \underline{P}(B) > 0. \end{cases}$$

Conditioning a credal set $\mathcal{M}$ corresponds to taking the credal set $\mathcal{M}|B$ formed by conditioning each of the precise previsions in $\mathcal{M}$:

$$\mathcal{M}|B = \begin{cases} \mathcal{P}_B & \text{if } \exists P \in \mathcal{M} : P(B) = 0, \\ \{P(\cdot \mid B) : P \in \mathcal{M}\} & \text{if } \forall P \in \mathcal{M} : P(B) > 0. \end{cases}$$

These rules based on natural extension give vacuous conditionals whenever the lower probability of the conditioning event is zero. *Regular extension* is a less imprecise updating rule [17, §2.3.4], [18, §1.6.6], [20, App. J]: In credal set terms, it removes those precise previsions P such that $P(B) = 0$ from $\mathcal{M}$.

Let us apply the conditioning rules discussed here to our running example:

Conditioning the empty-uniform pool's credal set

We condition the empty-uniform pool's credal set on $\{\text{JP}, \text{P}\}$, i.e., Belgium not winning in regular time. Further down on the left, we show what happens if we apply natural extension: the conditional model is vacuous because $P_{(1,0,0)}(\{\text{JP}, \text{P}\}) = P_{(1,0,0)}(1_{\{\text{JP},\text{P}\}}) = (1, 0, 0)^\top(0, 1, 1) = 0$. Further down on the right, we apply natural extension and therefore remove $P_{(1,0,0)}$ from $\mathcal{M}_{\text{EUP}}$; this results in a non-vacuous conditional credal set.

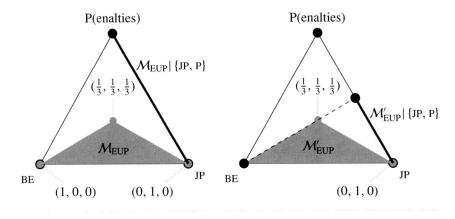

3.6 Remarks About Infinite Possibility Spaces

The theory we presented is also applicable to denumerable and continuous possibility spaces with some technical amendments to the coherence criteria and by considering only bounded gambles. However, the running example was based on finite possibility spaces, so no feeling was created for applications with infinite possibility spaces. Therefore we here give some remarks about imprecise probabilistic uncertainty models on continuous possibility spaces:

- They are mostly defined using credal sets whose extreme points are parametric distributions where the parameters vary in a set. A prime example are the imprecise Dirichlet model [21] and its generalizations [19].
- They are also commonly defined using probability mass assignments to subsets of the possibility space. This is in some way a reduction to the finite case. Examples are belief functions [13, §5.2.1.1], some P-boxes [14, §4.6.4], and NPI models [5, §7.6].
- Furthermore, models which bound some specific description of a precise prevision, such as cumulative distribution functions and probability density functions, are also popular in some domains. The extreme points of their credal set are, however, not known. General P-boxes [15] and lower and upper density functions [20, §4.6.3] are examples of this class.
- Calculating lower and upper previsions—i.e., performing natural extension—can easily become difficult optimization problems, so this should be a key consideration when choosing a specific type of model.

3.7 Conclusion

This introduction to the theory of imprecise probability has prepared you for accessing the broader literature on this topic and its applications. For those that wish to apply imprecise probabilistic techniques, this text only provides the first step: You should dive into the literature and contact experts to obtain the necessary knowledge and feedback. The references of this chapter and their authors or editors provide a starting point for that.

Acknowledgements I would like to thank Frank Coolen for giving me the opportunity to give a lecture at the 2018 UTOPIAE Training School and all the students for their active participation. Thanks are also due to the Wind Energy group at TU Delft, which gave me the freedom to contribute to the Training School, despite the demands of my regular duties.

References

1. Alessandro Antonucci, Giorgio Corani, Inés Couso, and Sébastien Destercke, editors. *ISIPTA '17: Proceedings of the Tenth International Symposium on Imprecise Probability: Theories and Applications*, volume 62 of *Proceedings of Machine Learning Research*, 2017.
2. Thomas Augustin, Frank P. A. Coolen, Gert de Cooman, and Matthias C. M. Troffaes, editors. *Introduction to Imprecise Probabilities*. Wiley Series in Probability and Statistics. Wiley, 2014.
3. Thomas Augustin, Frank P. A. Coolen, Serafín Moral, and Matthias C. M. Troffaes, editors. *ISIPTA '09: Proceedings of the Sixth International Symposium on Imprecise Probability: Theories and Applications*, 2009.
4. Thomas Augustin, Serena Doria, Enrique Miranda, and Erik Quaeghebeur, editors. *ISIPTA '15: Proceedings of the Ninth International Symposium on Imprecise Probability: Theories and Applications*, 2015.
5. Thomas Augustin, Gero Walter, and Frank P. A. Coolen. Statistical inference. In Augustin et al. [2], chapter 7, pages 135–189.
6. Jean-Marc Bernard, Teddy Seidenfeld, and Marco Zaffalon, editors. *ISIPTA '03: Proceedings of the Third International Symposium on Imprecise Probabilities and Their Applications*, volume 18 of *Proceedings in Informatics*, Waterloo, Ontario, Canada, 2003. Carleton Scientific.
7. Frank P. A. Coolen, Gert de Cooman, Thomas Fetz, and Michael Oberguggenberger, editors. *ISIPTA '11: Proceedings of the Seventh International Symposium on Imprecise Probability: Theories and Applications*, 2011.
8. Fabio Cozman, Thierry Denœux, Sébastien Destercke, and Teddy Seidenfeld, editors. *ISIPTA '13: Proceedings of the Eight International Symposium on Imprecise Probability: Theories and Applications*, 2013.
9. Fabio Gagliardi Cozman, Robert Nau, and Teddy Seidenfeld, editors. *ISIPTA '05: Proceedings of the Fourth International Symposium on Imprecise Probabilities and Their Applications*, 2005.
10. Gert de Cooman, Fabio Gagliardi Cozman, Serafín Moral, and Peter Walley, editors. *ISIPTA '99: Proceedings of the First International Symposium on Imprecise Probabilities and Their Applications*, 1999.
11. Gert de Cooman, Terrence L. Fine, and Teddy Seidenfeld, editors. *ISIPTA '01: Proceedings of the Second International Symposium on Imprecise Probabilities and Their Applications*, Maastricht, the Netherlands, 2001. Shaker Publishing.

12. Gert de Cooman, Jiřina Vejnarová, and Marco Zaffalon, editors. *Proceedings of the Fifth International Symposium on Imprecise Probabilities: Theories and Applications*. SIPTA, Action M Agency for SIPTA, 2007.
13. Sébastien Destercke and Didier Dubois. Other uncertainty theories based on capacities. In Augustin et al. [2], chapter 5, pages 93–113.
14. Sébastien Destercke and Didier Dubois. Special cases. In Augustin et al. [2], chapter 4, pages 79–92.
15. Scott Ferson, Vladik Kreinovich, Lev Ginzburg, Davis S. Myers, and Kari Sentz. Constructing probability boxes and dempster-shafer structures. Technical Report SAND2002-4015, Sandia National Laboratories, 2002.
16. Enrique Miranda. A survey of the theory of coherent lower previsions. *International Journal of Approximate Reasoning*, 48(2):628–658, 2008.
17. Enrique Miranda and Gert de Cooman. Lower previsions. In Augustin et al. [2], chapter 2, pages 28–55.
18. Erik Quaeghebeur. Desirability. In Augustin et al. [2], chapter 1, pages 1–27.
19. Erik Quaeghebeur and Gert de Cooman. Imprecise probability models for inference in exponential families. In Cozman et al. [9], pages 287–296.
20. Peter Walley. *Statistical reasoning with imprecise probabilities*, volume 42 of *Monographs on Statistics and Applied Probability*. Chapman & Hall, 1991.
21. Peter Walley. Inferences from multinomial data: learning about a bag of marbles. *Journal of the Royal Statistical Society B: Methodological*, 58(1):3–57, 1996.

Chapter 4
Imprecise Discrete-Time Markov Chains

Gert de Cooman

Abstract I present a short and easy introduction to a number of basic definitions and important results from the theory of imprecise Markov chains in discrete time, with a finite state space. The approach is intuitive and graphical.

4.1 Introduction

Although imprecision and robustness in discrete-time Markov chains were already studied in the 1990s [6–8], more significant progress [2, 3, 5, 11] could be made after the graphical structure of imprecise probability trees underlying them was uncovered in 2008 [4]. Research has now moved firmly into the continuous-time domain, for which [1, 9] are good starting points.

In this paper, I give a concise and elementary overview of a number of basic ideas and results in discrete-time imprecise Markov chains, with an emphasis on their graphical representation. We begin with the basics of precise and imprecise probability models in Sects. 4.2 and 4.3. When such models are used in a dynamical context, precise and imprecise probability trees arise naturally; they and the use of the fundamental Law of Iterated Expectations for making inferences about them constitute the subjects of Sects. 4.4 and 4.5. Imprecise Markov chains correspond to special imprecise probability trees, and they and their basic inferences are discussed in Sect. 4.6, followed by a number of examples in Sect. 4.7. These examples hint at stationary distributions and ergodicity. These notions are briefly discussed in Sect. 4.8, which concludes the paper. Throughout, I have included a number of simple exercises to illustrate the arguments in the main text.

G. de Cooman (✉)
Foundations Lab for imprecise probabilities, Ghent University, Ghent, Belgium
e-mail: gert.decooman@ugent.be

© The Author(s) 2022
L. Aslett et al. (eds.), *Uncertainty in Engineering*,
SpringerBriefs in Statistics,
https://doi.org/10.1007/978-3-030-83640-5_4

4.2 Precise Probability Models

Assume we are uncertain about the value that a variable X assumes in some finite set of possible values $\mathcal{X}$. This is usually modelled by a *probability mass function m* on $\mathcal{X}$, satisfying $(\forall x \in \mathcal{X})m(x) \geq 0$ and $\sum_{x \in \mathcal{X}} m(x) = 1$.

With m we can associate an *expectation operator* E_m as follows

$$E_m(f) := \sum_{x \in \mathcal{X}} m(x)f(x) \text{ where } f: \mathcal{X} \to \mathbb{R}.$$

If $A \subseteq \mathcal{X}$ is an *event*, then its *probability* is given by $P_m(A) = \sum_{x \in A} m(x) = E_m(I_A)$, where $I_A: \mathcal{X} \to \mathbb{R}$ is the *indicator* of A and assumes the value 1 on A and 0 elsewhere. This tells us that there are two equivalent mathematical languages for dealing with uncertainty: the language of probabilities and the language of expectations, and that we can go freely from one to the other.

All possible (precise) probability models are gathered in the *simplex* $\Sigma_\mathcal{X}$ of all mass functions on $\mathcal{X}$: $\Sigma_\mathcal{X} := \{m \in \mathbb{R}^\mathcal{X} : (\forall x \in \mathcal{X})m(x) \geq 0 \text{ and } \sum_{x \in \mathcal{X}} m(x) = 1\}$. Any probability model for uncertainty about X is a point in that simplex, which indicated that mass functions have a geometrical interpretation. This is illustrated below for the case $\mathcal{X} = \{a, b, c\}$ and the uniform mass function m_u.

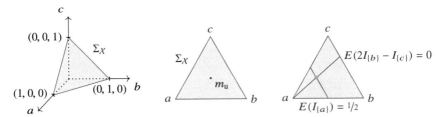

Expectation also has a geometrical interpretation: specifying a value $E(f)$ for the expectation of a map $f: \mathcal{X} \to \mathbb{R}$, namely, $\sum_{x \in \mathcal{X}} m(x)f(x) = E(f)$, imposes a linear constraint on the possible values for m in $\Sigma_\mathcal{X}$. It corresponds to intersecting the simplex $\Sigma_\mathcal{X}$ with a hyperplane, whose direction depends on f. This is also illustrated in the picture above; in this particular case two assessments turn out to completely determine a unique mass function.

4.3 Imprecise Probability Models

We now turn to a generalisation of precise probability models, which we will call imprecise. To allow for more realistic and flexible assessments, we can envisage imposing linear inequality—rather than equality—constraints on the m in $\Sigma_\mathcal{X}$:

$$\underline{E}(f) \leq \sum_{x \in \mathcal{X}} m(x)f(x) \text{ or } \sum_{x \in \mathcal{X}} m(x)f(x) \leq \overline{E}(f).$$

This corresponds to intersecting $\Sigma_{\mathcal{X}}$ with affine semi-spaces:

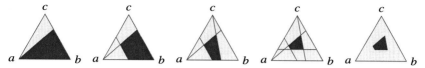

Any such number of assessments leads to a *credal set* $\mathcal{M}$, which is our first type of imprecise probability model.

Definition 4.1 A credal set $\mathcal{M}$ is a convex closed subset of $\Sigma_{\mathcal{X}}$.

Below, we show some more examples of such credal sets in the special case $\mathcal{X} = \{a, b, c\}$. The credal set on the left corresponds to the assessment: 'b is at least as likely as c'; the one in the middle is a convex mixture of the uniform mass function with the entire simplex; and the one on the right represents a statement in classical propositional logic: '$X = a$ or $X = c$'. This illustrates that the language of credal sets encompasses both precise probabilities and classical propositional logic.

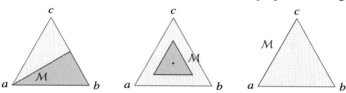

Lower and upper expectations are our second type of imprecise probability model. To see how they come about, consider the credal set in the figure below on the right.

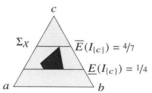

We can ask what we know about the probability of c, or the expectation of $I_{\{c\}}$, given this credal set: it is only known to belong to the closed interval $[1/4, 4/7]$. This can be generalised from events to arbitrary elements of the set $\mathcal{L}(\mathcal{X}) = \mathbb{R}^{\mathcal{X}}$ of all real-valued maps f on $\mathcal{X}$: As m ranges over the credal set $\mathcal{M}$, $E_m(f)$ will similarly range over a closed interval that is completely determined by its lower and upper bounds.

This leads to the definition of the following two real functionals on $\mathcal{L}(\mathcal{X})$:

$$\underline{E}_{\mathcal{M}}(f) = \min\{E_m(f) : m \in \mathcal{M}\} \text{ lower expectation}$$
$$\overline{E}_{\mathcal{M}}(f) = \max\{E_m(f) : m \in \mathcal{M}\} \text{ upper expectation} \quad \text{for all } f : \mathcal{X} \to \mathbb{R}.$$

Observe that these lower and upper expectations are mathematically equivalent models, because

$$\overline{E}_{\mathcal{M}}(f) = -\underline{E}_{\mathcal{M}}(-f) \text{ for all } f \in \mathcal{L}(\mathcal{X}).$$

We will in what follows focus on upper expectations.

Exercise 4.1 What is the upper expectation $\overline{E}_{\mathcal{M}}$ when $\mathcal{M} = \Sigma_{\mathcal{X}}$?

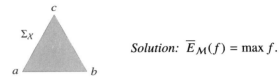

Solution: $\overline{E}_{\mathcal{M}}(f) = \max f$.

This shows that we can go from the language of probabilities—and the use of $\mathcal{M}$—to the language of expectations—and the use of $\overline{E}_{\mathcal{M}}$. To see that we can also go the other way, we need the following definition:

Definition 4.2 We call a real functional $\overline{E}$ on $\mathcal{L}(\mathcal{X})$ an *upper expectation* if it satisfies the following properties: for all f and g in $\mathcal{L}(\mathcal{X})$ and all real $\lambda \geq 0$:

1. $\overline{E}(f) \leq \max f$ [*boundedness*];
2. $\overline{E}(f + g) \leq \overline{E}(f) + \overline{E}(g)$ [*sub-additivity*];
3. $\overline{E}(\lambda f) = \lambda \overline{E}(f)$ [*non-negative homogeneity*].

Upper expectations are also called *coherent upper previsions* [10, 12]. They constitute a model that is mathematically equivalent to credal sets, in very much the same way as expectations are mathematically equivalent to probability mass functions:

Theorem 4.1 *A real functional $\overline{E}$ is an upper expectation if and only if it is the upper envelope of some credal set $\mathcal{M}$.*

Proof Use $\mathcal{M} = \left\{ m \in \Sigma_{\mathcal{X}} : (\forall f \in \mathcal{L}(\mathcal{X}))(E_m(f) \leq \overline{E}(f)) \right\}$. $\qquad\square$

Exercise 4.2 Consider any linear prevision E_m and any $\epsilon \in [0, 1]$. Verify that the so-called *linear-vacuous mixture*:

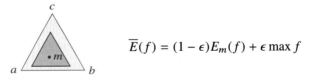

$$\overline{E}(f) = (1 - \epsilon)E_m(f) + \epsilon \max f$$

is an upper expectation.

Solution: E_m and max are upper expectations by Theorem 4.1, because they are upper envelopes of the respective credal sets $\{m\}$ and $\Sigma_{\mathcal{X}}$—see Exercise 4.1. Now verify that being an upper expectation is preserved by taking convex mixtures. The corresponding credal set $(1 - \epsilon)\{m\} + \epsilon \Sigma_{\mathcal{X}} := \{(1 - \epsilon)m + \epsilon q : q \in \Sigma_{\mathcal{X}}\}$ is indicated in blue in the figure above. $\qquad\diamond$

Exercise 4.3 All upper expectations on a binary space $\mathcal{X} = \{0, 1\}$ are such linear-vacuous mixtures, and the corresponding credal set can be depicted as

Let $p := m(1)$ and $q := 1 - p = m(0)$. What is the relation between $[\underline{p}, \overline{p}]$ and p, ϵ?

Solution: $\underline{p} = \underline{E}(I_{\{1\}}) = (1 - \epsilon)p = p - \epsilon p$ and $\overline{p} = \overline{E}(I_{\{1\}}) = (1 - \epsilon)p + \epsilon = p + \epsilon q$. Hence, $\overline{p} - \underline{p} = \epsilon$. $\Diamond$

4.4 Discrete-Time Uncertain Processes

We now apply these ideas in a more dynamic context: the study of processes. We consider an uncertain process, which is a collection of uncertain variables $X_1, X_2, \dots, X_n, \dots$ assuming values in some finite set of states $\mathcal{X}$. This can be represented graphically by a standard *event tree* with *nodes* (also called *situations*) $s = (x_1, x_2, \dots, x_n)$ for $x_k \in \mathcal{X}$ and $n \geq 0$. This is depicted below on the left for the special case that $\mathcal{X} = \{0, 1\}$, where we have limited ourselves to three variables $X_1, X_2,$ and X_3; but the idea should be clear. Observe that we use the symbol $\square$ for the *initial situation*, or root node, of the event tree.

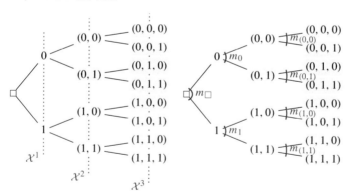

The event tree becomes a *probability tree* as soon as we attach to each node $s = (x_1, x_2, \dots, x_n)$ a *local* probability mass function m_s on $\mathcal{X}$ with associated expectation operator E_{m_s}, expressing the uncertainty about the next variable X_{n+1} after observing the earlier variables $X_1 = x_1, \dots, X_n = x_n$. This is depicted above on the right for the special case that $\mathcal{X} = \{0, 1\}$.

We now consider a very general inference problem in such a probability tree. Consider any function $g \colon \mathcal{X}^n \to \mathbb{R}$ of the first n variables: $g = g(X_1, X_2, \dots, X_n)$. We want to calculate its expectation $E(g|s)$ in the situation $s = (x_1, \dots, x_k)$, that is, after having observed the first k variables. Interestingly, this can be done efficiently using the following theorem, which is a reformulation of the Law of Total Probability:

Theorem 4.2 (Law of Iterated Expectations) *If we know $E(g|s, x)$ for all $x \in \mathcal{X}$, then we can calculate $E(g|s)$ by backwards recursion using the local model m_s:*

$$E(g|s) = \underbrace{E_{m_s}}_{local}(E(g|s, \cdot)) = \sum_{x \in \mathcal{X}} m_s(x)E(g|s, x).$$

This shows that expectations can be calculated recursively using a very basic step, illustrated below for the case $\mathcal{X} = \{0, 1\}$:

$$E(g|s) = m_s(1)E(g|s, 1) + m_s(0)E(g|s, 0) \longleftarrow s \overset{(s, 1) \longleftarrow E(g|s, 1)}{\underset{(s, 0) \longleftarrow E(g|s, 0)}{m_s}}$$

Hence, all expectations $E(g|x_1, \ldots, x_k)$ in the tree can be calculated from the local models m_s as follows:

1. start in the final cut $\mathcal{X}^n$ and let $E(g|x_1, x_2, \ldots, x_n) = g(x_1, x_2, \ldots, x_n)$;
2. do backwards recursion using the Law of Iterated Expectations:

$$E(g|x_1, \ldots, x_k) = \underbrace{E_{m_{(x_1, \ldots, x_k)}}}_{local}(E(g|x_1, \ldots, x_k, \cdot))$$

3. go on until you get to the root node $\square$, where we can identify $E(g|\square) = E(g)$.

Exercise 4.4 Consider flipping a coin twice independently, with probability p for heads—outcome 1—and $q = 1 - p$ for tails—outcome 0. The corresponding probability tree for this experiment is given below on the left, with, in red, in the nodes, the corresponding number of heads. What is the expected number of heads?

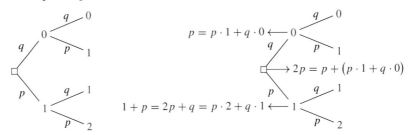

Solution: Above on the right, we apply the Law of Iterated Expectations recursively, from leaves to root; the solution is the expectation $2p$ attached to the root.   ◊

Exercise 4.5 Extend the ideas in the solution to Exercise 4.4 to calculate the expected number of heads when the coin is flipped n times independently.
Solution: We apply the Law of Iterated Expectations recursively, from leaves to root. Below on the left, we consider starting from the leaves of the tree at depth n; applying the Law reduces to adding p to the number of heads in each of their parent nodes at depth $n - 1$. On the right, we apply the Law to these nodes at depth $n - 1$, which reduces to adding $2p$ to the number of heads in each of their parent nodes at depth $n - 2$.

at time $n-1$ in a situation with k heads   at time $n-2$ in a situation with k heads

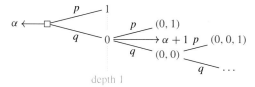

$$k \mathrel{\substack{q \\ \rightleftarrows \\ p}} \substack{k \\ k+1} \; k+p = k+(p\cdot 1 + q\cdot 0) \qquad k \mathrel{\substack{q \\ \rightleftarrows \\ p}} \substack{k+p \\ k+p+1} \; k+2p = k+p+(p\cdot 1 + q\cdot 0)$$

Going on in this way, we see that the solution is the expectation np attached to the root at depth 0.   ◇

Exercise 4.6 We now flip the same coin time and time again, independently, until we reach heads for the first time. Calculate the expected number of coin flips.
Solution: Below is the (unbounded) probability tree associated with this experiment.

$$\alpha \leftarrow \Box \mathrel{\substack{p \\ \nearrow \\ \searrow \\ q}} \substack{1 \\ 0} \mathrel{\substack{p \\ \rightarrow \\ q}} \substack{(0,1) \\ \alpha+1} \mathrel{\substack{p \\ \rightarrow \\ q}} \substack{(0,0,1) \\ (0,0)} \mathrel{\substack{p \\ \nearrow \\ \searrow \\ q}} \substack{(0,0,1) \\ \cdots}$$

depth 1

Call the unknown expectation α. We apply the Law of Iterated Expectations to the situations at depth 1. In the situation 1, the expected number of heads is 1, the actual number of heads there. In the situation 0, we see a copy of the original tree extending to the right, but since we have already flipped the coin once here, the expected number of heads in this situation is $\alpha + 1$. In the parent node, the expected number of heads α is therefore also given by $p \cdot 1 + q \cdot (\alpha + 1) = 1 + q\alpha$, whence $\alpha = 1/p$.   ◇

4.5 Imprecise Probability Trees

Until now, we have assumed that we have sufficient information in order to specify, in each node s, a local probability mass function m_s on the set $\mathcal{X}$ of possible values for the next state.

$$s \mathrel{\substack{\nearrow \\ m_s \\ \searrow}} \substack{(s,1) \\ (s,0)} \qquad \longrightarrow \qquad s \mathrel{\substack{\nearrow \\ \mathcal{M}_s \\ \searrow}} \substack{(s,1) \\ (s,0)}$$

We now let go of this major restrictive assumption by allowing for more general uncertainty models. We will consider credal sets as our more general local uncertainty models: closed convex subsets $\mathcal{M}_s$ of $\Sigma_{\mathcal{X}}$. See the figure below for a special case when $\mathcal{X} = \{0, 1\}$.

Definition 4.3 An *imprecise probability tree* is an event tree where in each node s the local uncertainty model is a credal set $\mathcal{M}_s$, or equivalently, its associated upper expectation $\overline{E}_s$, with $\overline{E}_s(f) := \max\{E_m(f) : m \in \mathcal{M}_s\}$ for all $f \in \mathcal{L}(\mathcal{X})$.

An imprecise probability tree can be interpreted as an infinity of *compatible* precise probability trees: choose in each node s a probability mass function m_s from the set $\mathcal{M}_s$.

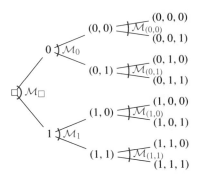

For each real map $g = g(X_1, \ldots, X_n)$, each node $s = (x_1, \ldots, x_k)$, and each such *compatible precise probability tree*, we can calculate the expectation $E(g|x_1, \ldots, x_k)$ using the backwards recursion method described before. By varying over each compatible probability tree, we get a closed real interval, completely characterised by lower and upper expectations $\underline{E}(g|x_1, \ldots x_k)$ and $\overline{E}(g|x_1, \ldots, x_k)$: $[\underline{E}(g|x_1, \ldots, x_k), \overline{E}(g|x_1, \ldots, x_k)]$. The complexity of calculating these bounds in this way is clearly exponential in the number of time steps n. But, there is a more efficient method to calculate them:

Theorem 4.3 (Law of Iterated Upper Expectations [4, 5]) *If we know $\overline{E}(g|s, x)$ for all $x \in \mathcal{X}$, then we can calculate $\overline{E}(g|s)$ by backwards recursion using the local model $\overline{E}_s$:*

$$\overline{E}(g|s) = \underbrace{\overline{E}_s}_{local} (\overline{E}(g|s, \cdot)) = \max_{m_s \in \mathcal{M}_s} \sum_{x \in \mathcal{X}} m_s(x)\, \overline{E}(g|s, x).$$

This shows that expectations can be calculated recursively using a very basic step, illustrated below for the case $\mathcal{X} = \{0, 1\}$:

$$\overline{E}(g|s) = \overline{E}_s(\overline{E}(g|s, \cdot)) \longleftarrow s \begin{array}{l} (s, 1) \longleftarrow \overline{E}(g|s, 1) \\ (s, 0) \longleftarrow \overline{E}(g|s, 0) \end{array}$$

The method for, and the complexity of, calculating the $\overline{E}(g|s)$, as a function of n, is therefore essentially the same as in the precise case!

Exercise 4.7 Extend the ideas in the solution to Exercise 4.5 to calculate the upper expected number of heads when the coin is flipped n times independently, but where now we have an imprecise probability model for a coin flip, with a probability interval $[\underline{p}, \overline{p}]$ for heads, and a corresponding interval $[\underline{q}, \overline{q}] = [1 - \overline{p}, 1 - \underline{p}]$ for tails.
Solution: We apply the Law of Iterated Upper Expectations recursively, from leaves to root. Below on the left, we consider starting from the leaves of the tree at depth n;

applying the Law reduces to adding $\overline{p}$ to the number of heads in each of their parent nodes at depth $n-1$.

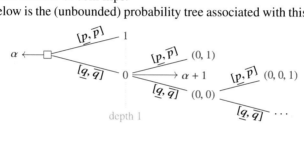

On the right, we apply the Law to these nodes at depth $n-1$, which reduces to adding $2\overline{p}$ to the number of heads in each of their parent nodes at depth $n-2$. Going on in this way, we see that the solution is the expectation $n\overline{p}$ attached to the root at depth 0. A similar result holds for the lower expectation. $\diamondsuit$

Exercise 4.8 We now flip the same coin with the imprecise probability model time and time again, independently, until we reach heads for the first time. Calculate the upper expected number of coin flips.

Solution: Below is the (unbounded) probability tree associated with this experiment.

Call the unknown upper expectation α. We apply the Law of Iterated Upper Expectations to the situations at depth 1. In the situation 1, the upper expected number of heads is 1, the actual number of heads there. In the situation 0, we see a copy of the original tree extending to the right, but since we have already flipped the coin once here, the upper expected number of heads in this situation is $\alpha + 1$. In the parent node, the upper expected number of heads α is therefore also given by $1 + \overline{E}(\alpha I_{\{0\}}) = 1 + \alpha\overline{q}$, whence $\alpha = 1/\underline{p}$. A similar result holds for the lower expectation. $\diamondsuit$

The attentive reader will have observed that in all these simple exercises, we can also obtain the 'imprecise' result from the 'precise' one by optimising over the single parameter p. We have to warn against too much optimism: in more involved examples, this will no longer be the case.

4.6 Imprecise Markov Chains

We now look at a special instance of a probability tree, corresponding to a stationary (precise) Markov chain. This happens when the precise local models $m_{(x_1,\ldots,x_n)}$ only depend on the last observed state x_n—this is the *Markov Condition*—and also do not depend explicitly on the time step n:

$$m_{(x_1,\dots,x_n)} = q(\cdot|x_n)$$

for some family of *transition mass functions* $q(\cdot|x)$, $x \in \mathcal{X}$.

Definition 4.4 The uncertain process is a stationary *precise Markov chain* when all $\mathcal{M}_s$ are singletons $\{m_s\}$ and $\mathcal{M}_{(x_1,\dots,x_n)} = \{q(\cdot|x_n)\}$, for some family of *transition mass functions* $q(\cdot|x)$, $x \in \mathcal{X}$.

For each $x \in \mathcal{X}$, the transition mass function $q(\cdot|x)$ corresponds to an expectation operator, given by $E(f|x) = \sum_{z \in \mathcal{X}} q(z|x) f(z)$ for all $f \in \mathcal{L}(\mathcal{X})$.

Definition 4.5 Consider the linear transformation T of $\mathcal{L}(\mathcal{X})$, called *transition operator*: $T: \mathcal{L}(\mathcal{X}) \to \mathcal{L}(\mathcal{X})$: $f \mapsto Tf$, where Tf is the real map defined by:

$$Tf(x) := E(f|x) = \sum_{z \in \mathcal{X}} q(z|x) f(z) \text{ for all } x \in \mathcal{X}.$$

In the parlance of linear algebra, or functional analysis, T is the dual of the linear transformation with Markov matrix M with elements $M_{xy} := q(y|x)$.

Up to now, we have mainly been concerned with conditional expectations of the type $E(\cdot|s)$. We will now look at particular *unconditional* expectations, where $s = \square$. For any $n \geq 0$, we define the expectation for the (single) state X_n at time n by

$$E_n(f) = E(f(X_n)) = E(f(X_n)|\square) \text{ for all } f : \mathcal{X} \to \mathbb{R}$$

and we denote the corresponding mass function by m_n. Applying the Law of Iterated Expectations in Theorem 4.2 now yields, with also $E_1 = E_{m_\square}$ and $m_1 = m_\square$:

$$E_n(f) = E_1(T^{n-1} f), \text{ and dually, } m_n = M^{n-1} m_1,$$

so the complexity of calculating $E_n(f)$ is linear in the number of time steps n.

Exercise 4.9 Consider the stochastic process where we first flip a fair coin. From then on, on heads, we select a biased coin with probability p for heads for the next coin flip, and on tails, a biased coin with probability $q = 1 - p$ for heads, and keep on flipping one of the two biased coins, selected on the basis of the outcome of the previous coin flip. This produces a Markov chain. Find Tf, $T^2 f$, and $E_1(f)$, $E_2(f)$ and $E_3(f)$ for $f \in \mathcal{L}(\{0, 1\})$.
Solution: Clearly, $E_1(f) = \frac{1}{2} f(1) + \frac{1}{2} f(0)$, $Tf(0) = E(f|0) = qf(1) + pf(0)$ and $Tf(1) = E(f|0) = pf(1) + qf(0)$, whence

$$E_2(f) = E_1(Tf) = \frac{p+q}{2} f(1) + \frac{p+q}{2} f(0) = \frac{1}{2} f(1) + \frac{1}{2} f(0).$$

Similarly,

$$T^2 f(0) = qTf(1) + pTf(0) = q[pf(1) + qf(0)] + p[qf(1) + pf(0)]$$
$$= (p^2 + q^2)f(0) + 2pqf(1)$$
$$T^2 f(1) = (p^2 + q^2)f(1) + 2pqf(0),$$

whence

$$E_3(f) = E_1(T^2 f) = \frac{p^2 + q^2 + 2pq}{2} f(1) + \frac{2pq + p^2 + q^2}{2} f(0) = \frac{1}{2} f(1) + \frac{1}{2} f(0),$$

and so on. We see that at the level of expectations of single state variables, the process cannot be distinguished from flipping a fair coin. ◇

The generalisation from precise to imprecise Markov chains goes as follows:

Definition 4.6 The uncertain process is a stationary *imprecise Markov chain* when the Markov Condition is satisfied with stationarity: $\mathcal{M}_{(x_1,\dots,x_n)} = \mathcal{Q}(\cdot|x_n)$ for some family of *transition credal sets* $\mathcal{Q}(\cdot|x)$, $x \in \mathcal{X}$.

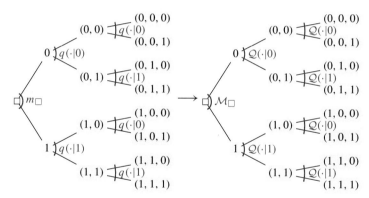

An imprecise Markov chain can be seen as an infinity of (precise) probability trees: choose a precise mass function from $\mathcal{M}_s$ in each situation s. It should be clear that not all of these satisfy the Markov property or stationarity. This implies that solving the optimisation problem in order to find the tight upper bounds $\overline{E}(g|s)$, as discussed in Sect. 4.5, is not (necessary always) simply an optimisation over a parametrised collection of stationary (or even non-stationary) Markov chains, although it can turn out be so simple in a number of special cases.

For each $x \in \mathcal{X}$, the local transition model $\mathcal{Q}(\cdot|x)$ corresponds to an upper expectation operator $\overline{E}(\cdot|x)$, with $\overline{E}(f|x) = \max\{E_p(f) : p \in \mathcal{Q}(\cdot|x)\}$ for all $f \in \mathcal{L}(\mathcal{X})$. This leads to the following definition, which generalises the definition of transition operators for precise Markov chains:

Definition 4.7 Consider the non-linear transformation $\overline{T}$ of $\mathcal{L}(\mathcal{X})$, called the *upper transition operator:* $\overline{T} \colon \mathcal{L}(\mathcal{X}) \to \mathcal{L}(\mathcal{X}) \colon f \mapsto \overline{T}f$ where the real map $\overline{T}f$ is defined by $\overline{T}f(x) := \overline{E}(f|x) = \max\{E_p(f) : p \in \mathcal{Q}(\cdot|x)\}$ for all $x \in \mathcal{X}$.

For any $n \geq 0$, we define the upper expectation for the (single) state X_n at time n by

$$\overline{E}_n(f) = \overline{E}(f(X_n)) = \overline{E}(f(X_n)|\square) \text{ for all } f : \mathcal{X} \to \mathbb{R}.$$

Then the Law of Iterated Upper Expectations of Theorem 4.3 yields, with also $\overline{E}_1 = \overline{E}_{\mathcal{M}_\square}$:

$$\overline{E}_n(f) = \overline{E}_1(\overline{T}^{n-1} f) \text{ for all } n \geq 1 \text{ and all } f \in \mathcal{L}(\mathcal{X}),$$

so the complexity of calculating $\overline{E}_n(f)$ is still linear in the number of time steps n.

4.7 Examples

Consider a two-element state space $\mathcal{X} = \{1, 0\}$, with upper expectation $\overline{E}_1 = \overline{E}_{\mathcal{M}_\square}$ for the first variable, and for each $(x_1, \ldots, x_n) \in \{1, 0\}^n$, with $0 < \epsilon \leq 1$, $\mathcal{M}_{(x_1,\ldots,x_n)} = \mathcal{M}_{x_n} = (1 - \epsilon)\{q(\cdot|x_n)\} + \epsilon \Sigma_{\{1,0\}}$, or equivalently, for the upper transition operator $\overline{T} = (1 - \epsilon)T + \epsilon \max$. In other words, each transition credal set $\mathcal{Q}(\cdot|x)$ is a linear-vacuous mixture (see Exercise 4.2, also for the notations used) centred on the transition mass function $q(\cdot|x)$, where the mixture coefficient ϵ is the same in each state x.

It is a matter of simple and direct verification that for $n \geq 1$ and $f \in \mathcal{L}(\mathcal{X})$: $\overline{T}^n f = (1 - \epsilon)^n T^n f + \epsilon \sum_{k=0}^{n-1}(1 - \epsilon)^k \max T^k f$, and therefore, using the Law of Iterated Expectations, $\overline{E}_{n+1}(f) = \overline{E}_1(\overline{T}^n f) = (1 - \epsilon)^n \overline{E}_1(T^n f) + \epsilon \sum_{k=0}^{n-1}(1 - \epsilon)^k \max T^k f$. If we now let $n \to \infty$, it is not too hard to see that the limit exists and is independent of the initial upper expectation $\overline{E}_1$:

$$\lim_{n \to \infty} \overline{E}_n(f) = \epsilon \sum_{k=0}^{\infty}(1 - \epsilon)^k \max T^k f \text{ for all } f \in \mathcal{L}(\mathcal{X}).$$

We consider two special cases:

1. Contaminated random walk: when $Tf(1) = Tf(0) = 1/2[f(1) + f(0)]$, the underlying precise Markov chain is actually like flipping a fair coin. We then find that $\overline{E}_\infty(f) = (1 - \epsilon)1/2[f(1) + f(0)] + \epsilon \max f$ for all $f \in \mathcal{L}(\mathcal{X})$.
2. Contaminated cycle: when $Tf(1) = f(0)$ and $Tf(0) = f(1)$, the underlying precise Markov chain is actually like deterministic cycle between the states 0 and 1. We then find that $\overline{E}_\infty(f) = \max f$ for all $f \in \mathcal{L}(\mathcal{X})$.

The probability intervals for 1 corresponding to these two limit models are given by

As another example, we consider $\mathcal{X} = \{a, b, c\}$ and the transition models depicted below, which are imprecise models not very far from a simple cycle:

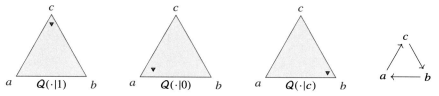

Below, we depict the time evolution of the $\overline{E}_n$ (as credal sets) for three cases (red, yellow and blue). We see that, here too, regardless of the initial distribution $\overline{E}_1$, the distribution $\overline{E}_n$ seems to converge to the same distribution.

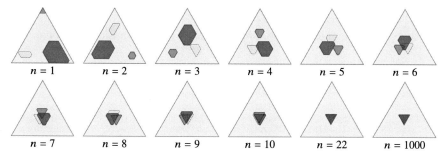

4.8 A Non-linear Perron–Frobenius Theorem, and Ergodicity

The convergence behaviour in the previous examples can also be observed in general imprecise Markov chains under fairly weak conditions. The following theorems can be derived from the more general discussions and results in [3, 5].

Theorem 4.4 *Consider a stationary imprecise Markov chain with finite state set* $\mathcal{X}$ *and upper transition operator* $\overline{T}$. *Suppose that* $\overline{T}$ *is regular, meaning that there is some* $n > 0$ *such that* $\min \overline{T}^n I_{\{x\}} > 0$ *for all* $x \in \mathcal{X}$. *Then for every initial upper expectation* $\overline{E}_1$, *the upper expectation* $\overline{E}_n = \overline{E}_1 \circ \overline{T}^{n-1}$ *for the state at time n converges point-wise to the same stationary upper expectation* $\overline{E}_\infty$: $\lim_{n \to \infty} \overline{E}_n(h) = \lim_{n \to \infty} \overline{E}_1(\overline{T}^{n-1} h) := \overline{E}_\infty(h)$ *for all h in* $\mathcal{L}(\mathcal{X})$. *The limit upper expectation* $\overline{E}_\infty$ *is the only* $\overline{T}$-invariant upper expectation on $\mathcal{L}(\mathcal{X})$, *meaning that* $\overline{E}_\infty = \overline{E}_\infty \circ \overline{T}$.

In that case we also have an interesting *ergodicity* result. For a detailed description of the notion of 'almost surely', we refer to [3], but it roughly means 'with upper probability one'.

Theorem 4.5 *Consider a stationary imprecise Markov chain with finite state set* $\mathcal{X}$ *and upper transition operator* $\overline{T}$. *Suppose that* $\overline{T}$ *is regular with stationary upper expectation* $\overline{E}_\infty$. *Then, almost surely, for all h in* $\mathcal{L}(\mathcal{X})$:

$$\underline{E}_\infty(h) \leq \liminf_{n\to\infty} \frac{1}{n} \sum_{k=1}^{n} h(X_k) \leq \limsup_{n\to\infty} \frac{1}{n} \sum_{k=1}^{n} h(X_k) \leq \overline{E}_\infty(h).$$

4.9 Conclusion

The discussion in this paper lays bare a few interesting but quite basic aspects of inference in imprecise probability trees and Markov chains in discrete time. A more general and deeper treatment of these matters can be found in [3–5]. For recent work on imprecise Markov chains in continuous time, I refer the interested reader to [1, 9].

References

1. Jasper De Bock. The Limit Behaviour of Imprecise Continuous-Time Markov Chains. *Journal of Nonlinear Science*, 27(1):159–196, 2017.
2. Jasper De Bock and Gert de Cooman. An efficient algorithm for estimating state sequences in imprecise hidden Markov models. *Journal of Artificial Intelligence Research*, 50:189–233, 2014.
3. Gert de Cooman, Jasper De Bock, and Stavros Lopatatzidis. Imprecise stochastic processes in discrete time: global models, imprecise markov chains, and ergodic theorems. *International Journal Of Approximate Reasoning*, 76(C):18–46, 2016.
4. Gert de Cooman and Filip Hermans. Imprecise probability trees: Bridging two theories of imprecise probability. *Artificial Intelligence*, 172(11):1400–1427, 2008.
5. Gert de Cooman, Filip Hermans, and Erik Quaeghebeur. Imprecise Markov chains and their limit behaviour. *Probability in the Engineering and Informational Sciences*, 23(4):597–635, 2009. arXiv:0801.0980.
6. D. J. Hartfiel and E. Seneta. On the theory of Markov set-chains. *Advances in Applied Probability*, 26:947–964, 1994.
7. Darald J. Hartfiel. Sequential limits in Markov set-chains. *Journal of Applied Probability*, 28(4):910–913, 1991.
8. Darald J. Hartfiel. *Markov Set-Chains*. Number 1695 in Lecture Notes in Mathematics. Springer, Berlin, 1998.
9. Thomas Krak, Jasper De Bock, and Arno Siebes. Imprecise continuous-time Markov chains. *International Journal of Approximate Reasoning*. 88:452–528, 2017.
10. Matthias C. M. Troffaes and Gert de Cooman. *Lower Previsions*. Wiley, 2014.
11. Damjan Škulj and Robert Hable. Coefficients of ergodicity for Markov chains with uncertain parameters. *Metrika*, 76(1):107–133, 2013.
12. Peter Walley. *Statistical Reasoning with Imprecise Probabilities*. Chapman and Hall, London, 1991.

Chapter 5
Statistics with Imprecise Probabilities—A Short Survey

Thomas Augustin

Abstract This chapter aims at surveying and highlighting in an introductory way some challenges and big opportunities a paradigmatic shift to imprecise probabilities could induce in statistical modelling. Working with an informal understanding of imprecise probabilities, we discuss the concepts of model imprecision and data imprecision as the two main types of imprecision in statistical modelling. Then we provide a short survey of some major developments, methodological questions and applications of imprecise probabilistic models under model imprecision in the context of different inference schools and summarize some recent developments in the area of data imprecision.

5.1 Introduction

By promising powerful solutions to some of the deepest foundational problems of probability and statistics, imprecise probabilities offer great opportunities also for the area of statistical modelling. Still, in statistics, theories of imprecise probabilities live in the shadows, and admittedly the development of many of the imprecise probability-based methods is often in a comparatively early stage. Nevertheless, in all areas of statistics, the desire for a more comprehensive modelling of complex uncertainty had popped up again and again. The rather scattered work covers a huge variety of methods and topics, ranging from contributions to the methodological and philosophical foundations of inference to very concrete questions of applications.

T. Augustin (✉)
Department of Statistics, Ludwig-Maximilians-Universität München (LMU Munich),
Munich, Germany
e-mail: augustin@stat.uni-muenchen.de

© The Author(s) 2022
L. Aslett et al. (eds.), *Uncertainty in Engineering*,
SpringerBriefs in Statistics,
https://doi.org/10.1007/978-3-030-83640-5_5

The present chapter aims at providing a rough and informal survey of some of the major questions and developments.[1] It is structured as follows. Section 5.2 collects the basic concepts that are needed later on. Then, Sect. 5.3 looks at the major sources of imprecision in statistical modelling. We distinguish there between several types of data imprecision and of model imprecision. Section 5.4 focuses on the issue of model imprecision and discusses it from the angle of different inference schools. We put some emphasis on the (generalized) frequentist and Bayesian setting, but also briefly adopt other perspectives. In Sect. 5.5, approaches to handle so-called ontic and epistemic data imprecision, respectively, are surveyed. Section 5.6 is reserved for some concluding remarks.

5.2 Some Elementary Background on Imprecise Probabilities

In this section, we briefly summarize the concepts in the background. With respect to the basic notions and the technical framework for statistical inference, we refer to the first chapter of this book [41]. We rely on the same basic setting, where we use observations (data) on some underlying stochastic phenomenon[2] to learn characteristics of that mechanism, mathematically described by an unknown parameter of the underlaid probability model.

With respect to imprecise probabilities, a very rough and eclectic understanding shall be sufficient to read this chapter.[3] It is not necessary, for our aims here, to distinguish different approaches with respect to many technical details. Thus, in a rather inclusive manner, we subsume here under *imprecise probabilities* any approach that replaces in its modelling precise, traditional probabilities $p(\cdot)$ by *non-empty sets* $\mathcal{P}$ *of precise probabilities* as the basic modelling entity, including also all approaches that can be equivalently transferred into a set of precise probability. This comprises approaches directly working with sets of probabilities, like robust Bayes analysis (see Sect. 5.4.2), Kofler & Menges' linear partial information (e.g. [42]), Levi's approach to epistemology (e.g. [45]), as well as the whole bunch of corresponding approaches based on non-linear functionals and non-additive set-functions, covering lower and upper previsions in tradition of Walley's book [66], interval probabilities building on

---

[1] A longer survey discussing the state of the art of statistical inference with imprecise probabilities at that time is aimed at by [8].

[2] To avoid any commitment to a certain interpretation of probability, we use the term "stochastic phenomenon" as a kind of neutral, superordinate concept, in particular including approaches referring to random phenomena only as well as approaches addressing generally situations of epistemic uncertainty.

[3] An introduction into imprecise probabilities on an intermediate level is aimed at by [4]; see also [57] (in this volume) for an introduction, [15] for a survey by a philosopher and [9] for a review with an engineering background. Current developments in research on imprecise probabilities are mostly discussed at the biannual ISIPTA (International Symposium on Imprecise Probabilities: Theories and Applications) meetings; see [1, 5, 27] for the most recent ones.

Weichselberger ([72][4]), probabilistically interpretable approaches based on capacities including the suitable branch of Dempster-Shafer theory (e.g. [29]), random sets (e.g. [12, Chap. 3]) and p-boxes following [32]. Moreover, there is a smooth transition to several approaches propagating systematic sensitivity analysis (e.g. [52]).

It is important that our basic entity, the set $\mathcal{P}$, has to be understood and treated as an entity of its own. It is by no means possible to distinguish certain of its elements as more likely than others or to mix its elements, eventually leading to a precise traditional probability distribution. $\mathcal{P}$ may be assessed directly by collecting several precise models, so-to-say as possible worlds/expert opinions, to be considered. More often $\mathcal{P}$ is constructed, typically as the set of all probability distributions

- that respect bounds on the probabilities of certain classes of events or, more generally, the expectations of certain random variables,[5, 6]
- or that are (in a topologically well-defined sense) close to a certain precise probability distribution $p_0(\cdot)$ (*neighbourhood models*), providing a formal framework for expressing statements like "$p_0(\cdot)$ is approximately true",[7]
- or that are described by a parameter ϑ varying in an interval/cuboid (*parametrically constructed models*), like "imprecise versions" of a normal distribution.[8]

5.3 Types of Imprecision in Statistical Modelling

There are many situations in statistics where imprecision occurs, i.e. where a careful modelling should go beyond the idealized situation of perfect stochasticity and data observed without any error and in an ideal precision. To study these situations further, an ideal-typical distinction between model and data precision is helpful.

Model Imprecision has to be taken into account whenever there is doubt in a concrete application that the strong requirements (perfect precision and, by the additivity axiom, absolute internal consistency) the traditional concept of probability calls for can be realistically satisfied. This includes all situations of incomplete or conflicting information, robustness concerns, and repeated sampling from a population where the common assumption of i.i.d. repetitions is violated by some unobserved heterogeneity or hidden dependence. In a Bayesian setting, in addition, quite often, if at all, a precise prior distribution is not honestly deducible from the knowledge actually available. Ellsberg's [31] seminal thought experiments on urns with only partially known compositions have made it crystal clear that the amount of ambiguity, i.e. the

---

[4] See also [73] for a summary in English language.

[5] For instance, in the approaches founded by [66, 72], respectively.

[6] It may be explicitly noted that the latter also includes comparative/ordinal probabilities (e.g. [49]), only ordering the probability of certain events like "A is more likely to occur than B, and B more likely than C".

[7] See Sect. 5.4.1.

[8] See, for instance, the sets of conjugated distributions in Sect. 5.4.2 for an example.

uncertainty about the underlying stochastic mechanism, plays a constitutive role in decision making and thus has to modelled carefully.

Data Imprecision comprises all situations where the observations are not available in the granularity that is originally intended in the corresponding application. Following [26], for the modelling of data imprecision, it is crucial to distinguish two situations, the precise observation of something inherently imprecise (*ontic data imprecision*) and the imprecise observation of something precise (*epistemic data imprecision*).[9] The difference between these two may be explained by a pre-election study where some voters are still undecided which single party they will vote for. (Compare [55], and for more details [44, 53]). If we understand a voting preference like "I am still undecided between parties *A* and *B*" as a political position of its own, then we interpret the information as an instance of ontic imprecision. If we focus on the forthcoming election and take this information as an imprecise observation allowing us to predict that the voting decision on the election day will either be *A* or *B*, then we are coping with epistemic imprecision. In particular in engineering, another frequent example of epistemic data imprecision occurs from insufficient measurement precision, where intervals instead of precise values are observed. A contrasting example where (unions of) intervals are to be understood as an entity of their own arises when the time span of certain spells is characterized: "This machine was under full load from November 10th to December 23rd."

Epistemic imprecision very naturally occurs in many studies on dynamic processes, from socio-economic over medical to technical studies. There, censoring is always a big issue: the spells of some units are still unfinished when the study ends, providing only lower bounds on the spell duration. Typical examples include the duration of unemployment, the time to recurrence of a tumour or the lifetime of an electronic component. In addition, also interval censoring is quite common, where one only learns that the event of interest occurred within a certain time span. It should be noted explicitly that missing data, a frequent problem for instance in almost every social survey, may be comprised under this setting, taking the whole sample space as the observation for those units missing.

5.4 Statistical Modelling Under Model Imprecision

Of course, there are also strong reservations against imprecise probabilities in traditional statistics. For a traditional statistician, imprecise probabilities are just a superfluous complication, misunderstanding either the generality of the concept of uncertainty or the reductionist essence of the modelling/abstraction process. Indeed, for an orthodox Bayesian, all kinds of not-knowing are simply situations under uncertainty, and any kind of uncertainty is eo ipso expressible by traditional probabilities. From the modelling perspective, imprecision is taken as part of the residual category

---

[9] See also the distinction between the disjunctive and conjunctive interpretation of random sets, as discussed, e.g. in [30, Sect. 1.4].

that is naturally lost when abstracting and building models.[10] Box (and Draper)'s often cited dictum "Essentially, all models are wrong, but some of them are useful" [14, p. 424] has generally been (mis)understood as a justification to base details of the model choice on mathematical convenience and as an irrevocable argument to take model imprecision as negligible.

5.4.1 Probabilistic Assumptions on the Sampling Model Matter: Frequentist Statistics and Imprecise Probabilities

Indeed, however, Box's quotation could be continued by "…and some models are dangerous", since the general neglecting of model imprecision implicitly presupposes a continuity (in an informal sense) of the conclusions in the models that by no means can be taken as granted. A well-known example from robust statistics[11] is statistical inference from a "regular bell-shaped distribution". The standard proceeding would be to assume a normal distribution. But, for instance, the density of a Cauchy distribution is phenomenologically de facto almost indistinguishable from the density function of a normal distribution, suggesting both models to be equivalent from a practical point of view. However, statistical inference based on the sample mean shows fundamentally different behaviours.[12] Under normality, the distribution of the sample mean behaves nicely, contracting itself around the correct value if the sample size n increases. In the case of the Cauchy distribution, however, the distribution of the sample mean stays the same irrespective of the sample size,[13] making any learning by the sample mean impossible.

This shocking insight—optimal statistical procedures may behave disastrously even under "tiny deviations" from the ideal model—demonstrates that imprecision in the underlying model may matter substantially. In this context, the theory of (frequentist) robust statistics as the theory of approximately true models emerged, and imprecise probabilities provide a natural superstructure upon it (see, e.g. [38] for historical connections). In particular, neighbourhood models, also briefly mentioned above, have become attractive.[14] Building on an influential result by Huber and Strassen [39], a comprehensive theory of testing in situations where the hypotheses

---

[10] This is particular the case when the prescriptive character of models is emphasized over its descriptive role (cf., e.g. [13, p. 99]).

[11] See also [8, Sect. 7.5.1, in particular, Fig. 7.3] to illustrate this.

[12] Generally, if one looks at an i.i.d. sample $X_1, \ldots, X_n$ from a normal distribution with parameters μ and σ^2, denoted by $\mathcal{N}(\mu, \sigma^2)$, and an i.i.d. sample $Z_1, \ldots, Z_n$ from a Cauchy distribution with parameter α and β, denoted by $C(\alpha, \beta)$, respectively, one obtains for the sample means $\bar{X} := \frac{1}{n} \sum_{i=1}^n X_i \sim \mathcal{N}(\mu, \frac{\sigma^2}{n})$ and $\bar{Z} := \frac{1}{n} \sum_{i=1}^n Z_i \sim C(\alpha, \beta)$.

[13] The Law of Large Numbers is not applicable to the Cauchy distribution because it does not have moments.

[14] See, e.g. [6] for a survey.

are described by imprecise probabilities emerged (see [2], [8, Sect. 7.5.2] and the references in the corresponding review sections therein.) Further insights are provided from interpreting frequentist statistics as a decision problem under an imprecise sampling distribution, leading to the framework investigated in [34].

Other frequentist approaches, starting from different angles, include Hampel's frequentist betting approach (e.g. [36]) and some work on minimum distance estimation under imprecise sampling models (e.g. [35]). The statistical consequences of the chaotic models in the genuine frequentist framework to imprecise probabilities, developed by Fine and followers (e.g. [33]), might be quite intriguing, but are still almost entirely unexplored.

5.4.2 Model Imprecision and Generalized Bayesian Inference

Priors and Sets of Priors, Generalized Bayes Rule. The centrepiece of Bayesian inference is the prior distribution. Apart from very large sample sizes, where the posterior is de facto determined by the sample, the prior naturally has a strong influence on the posterior and on all conclusions drawn from it. In the rare situations where very strong prior knowledge is available, it can be used actively, but most often the strong dependence on the prior has been intensively debated and criticized.

Working with sets Π of prior probabilities (or interval-valued priors) opens new avenues here. This set can naturally be chosen to reflect the quality/determinacy of prior knowledge: strong prior knowledge leads to "small" sets; weak prior knowledge to "large" sets. Typical model classes include neighbourhood models or sets of parametric distributions which often are conjugate [15] to the sampling distribution, which typically still is assumed to be precise.[16] In imprecise probability, Π is understood as naturally inducing the set Π_x of all posteriors arising from a prior in Π.[17] Interpretations of Π and Π_x vary to the extent they are understood as principled entities. A pragmatic point of view sees an investigation of Π_x just as a self-evident way to perform a sensitivity analysis. On the other extreme, Walley's [66] generalized theory of coherence, having initiated the most vivid branch of research on imprecise probabilities, provides a rigorous justification of exactly this way to proceed as the "Generalized Bayes Rule (GBR)". Important developments have also been achieved for a variety of different model classes under the term "Robust Bayesian Analysis"; see, e.g. [58] for a review on this topic.

Near Ignorance Models. One important way to use sets of priors is that they allow for quite a natural formulation of (rather) complete ignorance. A traditional Bayesian

---

[15] See, e.g. [41, Chap. 1.2] (in this volume).

[16] Imprecise sampling models are quite rarely studied, see, however, [66, Sect. 8.5] and [61].

[17] There are some approaches in the traditional Bayesian context that work with sets of priors, too, but merely understanding such a set as an interim step, based on which an ultimate precise prior is selected. This can be done in a data-driven way, as in empirical Bayes approaches, including the ML-II approach (e.g. [11, Sect. 3.5.4]), or generally, as in maximum entropy approaches (e.g. [11, Sect. 3.4]).

model eo ipso fails in expressing ignorance/non-informativeness. Assigning a precise probability is never non-committal; every precise prior delivers probabilistic information about the parameter. The genuinely non-informative model is the set of all probability distributions. While this model would yield vacuous inferences, it motivates so-called near-ignorance models, where, informally spoken, the inner core of this set is used, excluding extreme probabilities that are immune to learning. Near-ignorance models still assign non-committal probabilities to standard events in the parameter space, but allow for learning. By far the most popular model is the Imprecise Dirichlet Model (IDM) [67] for categorical data. Different extensions followed, including general near-ignorance models for exponential families (e.g. [10]). Another direction of enabling the formulation of near-ignorance uses all priors with bounded derivatives ([68]; for a general exposition of the concept of bounded influence, see the book [69]).

Prior-Data Conflict. In some sense, a complementary application of generalized Bayesian inference is the active modelling of prior-data conflict. In practice, generalized Bayesian models are quite powerful in expressing substantial prior knowledge. In particular, in areas where data are scarce, it is important to use explicitly all prior knowledge available, for instance by borrowing strength from similar experiments. Then, however, it is crucial to have some kind of alert system warning the analyst when the prior knowledge appears doubtful in the light of data. Indeed, sets of priors can be designed to react naturally to potential prior-data conflict: If data and prior assumptions are in agreement, the set of posterior distributions contracts more and more with increasing sample size. In contrast, in the case of prior-data conflict and intermediate sample size, the set of posterior distributions is inflated substantially, perfectly indicating that one should refrain from most decisions before having gathered further information.[18]

5.4.3 Some Other Approaches

With generalized Bayesian approaches, and less pronounced with generalized frequentist statistics, the major statistical inference school are also predominant in the area of imprecise probabilities. Nevertheless, there has also been considerable success in other inference frameworks. Again and again, the desire to save Fisher's fiducial argument, aiming at providing probability statements on parameters without having to rely on a prior distribution, has been a driving force for developments in imprecise probabilities. Dempster's concept of multivalued mappings (e.g. [28]), which become even more famous in artificial intelligence by Shafer's reinterpretation founding Dempster-Shafer Theory (see, for instance, again the survey by [29]), is to be mentioned here, but also work by Hampel (e.g. [36]) and by Weichselberger (e.g.

---

[18] See [71] for a general treatment in one-parameter exponential families, also providing some illustrating plots, and, for instance, [70] for an application in the context of system reliability.

[74]), see also [7] tracing back its roots.[19] A generalized likelihood-based framework has been introduced by Cattaneo (e.g. [16, 17]).

Another direct inference approach is *Nonparametric Predictive Inference (NPI)*, as introduced by Coolen originally for Bernoulli data [18]. Based on exchangeability arguments, NPI yields direct conditional probabilities of further real-valued random quantities, relying on the low structure assumption that all elements of the natural partition produced by the already observed data are equally likely; see, for instance, [19] for a detailed discussion, [3] for a clear embedding into imprecise probabilities and [21] for a web-page documenting research within this framework. The basic approach can be naturally extended to censored data/competing risks (e.g. [22]), and to categorical data [20]. NPI has been developed further in a huge variety of fields; see, for instance, [23, 24] for recent applications in biometrics and finance, respectively.

5.5   Statistical Modelling Under Data Imprecision

In this section, we turn to statistical modelling under data imprecision. Keeping the distinction from Sect. 5.3, we briefly discuss ontic data imprecision and then turn to epistemic data imprecision.

Ontic data imprecision, where we understand the imprecise observation as an entity of its own, may be argued to be a border case between classical statistics and its extensions. Technically, we change the sample space of each observation to (an appropriate subset of) the power set. For instance, recalling the election example from Sect. 5.3, this mean that instead of $\{a, b, c, \ldots\}$ representing the vote for a single party, we now also allow for combinations $\{a, b\}, \{b, c\}, \ldots, \{a, b, c\}, \ldots,$ representing the indecision between several parties. As long as we are in a multinomial setting, nothing has changed from an abstract point of view, providing powerful opportunities for complex statistical modelling. In the spirit of this idea, [44, 55] apply multinomial regression models, classification trees, regularized discrete choice models from election research, and spectral clustering methods to German pre-election survey data. The situation changes substantially when ordinal or continuous data are considered, because, after changing to the power set, the underlying ordering structure is only partially preserved.

Epistemic data imprecision is, as the examples at the end of Sect. 5.3 show, of great importance in many applications and is quite vividly addressed in classical statistics. Even here, traditional statistics keeps its focus on full identification, i.e. the selection of one single probability model fitting the observed data optimally. One searches for, and then implicitly relies on, conditions under which one gets hands on the so-to-say deficiency process as a thought pattern, making ideal precise observations imprecise. For that purpose, most classical approaches assume either some

---

[19]It may also be argued that, although less explicitly aiming at fiducial reasoning, the work on near-ignorance models discussed above fits well into this category.

kind of uninformativeness of the deficiency process (independent censorship, coarsening at random (CAR) or missingness (completely) at random (MCAR, MAR)) or an explicit modelling of the deficiency process; see the classical work by [37, 46]. Both the uninformativeness as well as the existence of a precisely specifiable deficiency process are very strong assumptions. They are—eo ipso by making explicit statements about unobservable processes—typically not empirically testable. Whenever these assumptions are just made for purely formal reasons, the price to pay for the seemingly precise result of the estimation process is high. In terms of Manski's *Law of Decreasing Credibility*,[20] results may suffer severely from a loss of their credibility, and thus of their practical relevance.

Against this background, in almost any area of application, the desire for less committal, cautious handling of epistemic data imprecision arose. Mostly isolated approaches were proposed that explicitly try to take all possible worlds into account in a reliable way, aiming at the *set* of all models optimally compatible with potentially true data. These approaches include, for instance, work from reliable computing and interval analysis in engineering, like [51], extensions of generalized Bayesian inference (e.g. [75]) to reliable statistics in social sciences (e.g. [56]); see also [8, Sect. 7.8.2], who try to characterize and unify these approaches by the concept of cautious data completion, and the concept of collection regions in [60].

There is a smooth transition to approaches that explicitly introduce cautious modelling into the construction of estimation procedures; see, for instance, for recent different likelihood- and loss minimization-based approaches addressing epistemic data imprecision, [25, 40, 43, 54]. Such approaches have the important advantage that their construction often also allows the incorporation of additional well-supported subject matter knowledge, too imprecise to be useful for the precision focused methods from traditional statistics, but very valuable to reduce the set of compatible models by a considerable extent.

Congenial is work in the field of partial identification and systematic sensitivity analysis, providing methodology for handling observationally equivalent models; see [48, 65], respectively, for classical work and [47, 62] for introductory surveys. The framework of partial identification is currently receiving considerable attention in econometrics, where in particular the embedding of fundamental questions into the framework of random sets is of particular importance [50].

5.6 Concluding Remarks

The contribution provided a—necessarily painfully selective—survey of some developments of statistical modelling with imprecise probabilities (in a wider sense, also including closely related concepts). Both in the area of model imprecision as well as under data imprecision, imprecise probabilities prove to be powerful and particularly

---

[20] "The credibility of inferences decreases with the strength of the assumptions maintained." [48, p. 1].

promising. Further developments urgently needed include a proper methodology for simulations with imprecise probabilities (see [64] for recent results), a careful study of the statistical consequences of the rather far developed probabilistic side of the theory of stochastic processes with imprecise probabilities (e.g. [63]), a more fruitful exchange with recent research on uncertainty quantification in engineering (see, e.g. [59] (in this volume)), an open mind towards recent developments in machine learning and more large scale applications. Not only for these topics it is important to complement the still recognizable focus on so-to-say defensive modelling by a more active modelling. Far beyond sensitivity and robustness aspects, imprecision can actively be used as a strong modelling tool. The proper handling of prior-data conflict and the successful incorporation of substantive matter knowledge in statistical analysis under data imprecision are powerful examples of going in this direction.

References

1. A. Antonucci, G. Corani, I. Couso, and S. Destercke, editors. *ISIPTA '17*, volume 62 of *Proceedings of Machine Learning Research*. PMLR, 2017.
2. T. Augustin. Neyman-Pearson testing under interval probability by globally least favorable pairs: Reviewing Huber-Strassen theory and extending it to general interval probability. *Journal of Statistical Planning and Inference*, 105:149–173, 2002.
3. T. Augustin and F. Coolen. Nonparametric predictive inference and interval probability. *Journal of Statistical Planning and Inference*, 124:251–272, 2004.
4. T. Augustin, F. P. A. Coolen, G. de Cooman, and M. Troffaes, editors. *Introduction to Imprecise Probabilities*. Wiley, Chichester, 2014.
5. T. Augustin, S. Doria, E. Miranda, and E. Quaeghebeur, editors. *ISIPTA '15*. SIPTA, 2015.
6. T. Augustin and R. Hable. On the impact of robust statistics on imprecise probability models: A review. *Structural Safety*, 32:358–365, 2010.
7. T. Augustin and Rudolf Seising. Kurt Weichselberger's contribution to imprecise probabilities. *International Journal of Approximate Reasoning*, 98:132–145, 2018.
8. T. Augustin, G. Walter, and F. Coolen. Statistical inference. In T. Augustin, F. Coolen, G. de Cooman, and M. Troffaes, editors, *Introduction to Imprecise Probabilities*, pages 135–189. Wiley, 2014.
9. M. Beer, S. Ferson, and V. Kreinovich. Imprecise probabilities in engineering analyses. *Mechanical Systems and Signal Processing*, 37:4–29, 2013.
10. A. Benavoli and M. Zaffalon. Prior near ignorance for inferences in the k-parameter exponential family. *Statistics*, 49:1104–1140, 2014.
11. J. O. Berger. *Statistical Decision Theory and Bayesian Analysis*. Springer, 2nd edition, 1985.
12. A. Bernardini and F. Tonon. *Bounding Uncertainty in Civil Engineering – Theoretical Background*. Springer, Berlin, 2010.
13. J. M. Bernardo and A. Smith. *Bayesian Theory*. Wiley, Chichester, 2000.
14. G. Box and N. Draper. *Empirical Model-building and Response Surface*. Wiley, New York, 1987.
15. S. Bradley. Imprecise probabilities. In Edward N. Zalta, editor, *The Stanford Encyclopedia of Philosophy (Spring 2019 Edition)*. Standford University, 2019. https://plato.stanford.edu/archives/spr2019/entries/imprecise-probabilities/.
16. M. Cattaneo. *Statistical Decisions Based Directly on the Likelihood Function*. PhD thesis, ETH Zurich, 2007.
17. M. Cattaneo. Likelihood decision functions. *Electronic Journal of Statistics*, 7:2924–2946, 2013.

18. F. Coolen. Low structure imprecise predictive inference for Bayes' problem. *Statistics & Probability Letters*, 36:349–357, 1998.
19. F. Coolen. On nonparametric predictive inference and objective Bayesianism. *Journal of Logic, Language and Information*, 15:21–47, 2006.
20. F. Coolen and T. Augustin. A nonparametric predictive alternative to the imprecise Dirichlet model: The case of a known number of categories. *International Journal of Approximate Reasoning*, 50:217–230, 2009.
21. F. Coolen, P. Coolen-Schrijner, and T. Coolen-Maturi. Nonparametric Predictive Inference (npi), 2020. https://npi-statistics.com/.
22. T. Coolen-Maturi. Nonparametric predictive pairwise comparison with competing risks. *Reliability Engineering & System Safety*, 132:146–153, 2014.
23. T. Coolen-Maturi. Predictive inference for best linear combination of biomarkers subject to limits of detection. *Statistics in Medicine*, 36:2844–2874, 2017.
24. T. Coolen-Maturi and F. Coolen. Nonparametric predictive inference for the validation of credit rating systems. *Journal of the Royal Statistical Society: Series A*, 182:1189–1204, 2019.
25. I. Couso and D. Dubois. A general framework for maximizing likelihood under incomplete data. *International Journal of Approximate Reasoning*, 93:238–260, 2018.
26. I. Couso, D. Dubois, and L. Sánchez. *Random Sets and Random Fuzzy Sets as Ill-Perceived Random Variables*. Springer, 2014.
27. J. De Bock, C.. de Campos, E. de Cooman, G.and Quaeghebeur, and G. Wheeler, editors. *ISIPTA '19*, volume 103 of *Proceedings of Machine Learning Research*. PMLR, 2019.
28. A. Dempster. Upper and lower probabilities induced by a multivalued mapping. *The Annals of Mathematical Statistics*, 38:325–339, 1967.
29. T. Denoeux. 40 years of Dempster-Shafer theory. *International Journal of Approximate Reasoning*, 79:1–6, 2016.
30. D. Dubois. Belief structure, possibility theory and decomposable confidence measures on finite sets. *Computers and Artificial Intelligence*, 5:403–416, 1986.
31. D. Ellsberg. Risk, ambiguity, and the Savage axioms. *Quarterly Journal of Economics*, 75:643–669, 1961.
32. S. Ferson, L. Ginzburg, V. Kreinovich, D. Myers, and K. Sentz. Constructing probability boxes and Dempster-Shafer structures (Sandia report), 2003. https://doi.org/10.2172/809606.
33. P. Fierens. An extension of chaotic probability models to real-valued variables. *International Journal of Approximate Reasoning*, 50:627–641, 2009.
34. R. Hable. *Data-based Decisions under Complex Uncertainty*. PhD thesis, LMU Munich, 2009.
35. R. Hable. Minimum distance estimation in imprecise probability models. *Journal of Statistical Planning and Inference*, 140:461–479, 2010.
36. F. Hampel. The proper fiducial argument. In R. Ahlswede, L. Bäumer, N. Cai, H. Aydinian, V. Blinovsky, C. Deppe, and H. Mashurian, editors, *General Theory of Information Transfer and Combinatorics*, volume 4123, pages 512–526. Springer, 2006.
37. D. Heitjan and D. Rubin. Ignorability and coarse data. *The Annals of Statistics*, 19:2244–2253, 1991.
38. P. Huber. The use of Choquet capacities in statistics. *Proceedings of the 39th Session of the International Statistical Institute*, 45:181–191, 1973.
39. P. Huber and V. Strassen. Minimax tests and the Neyman-Pearson lemma for capacities. *The Annals of Statistics*, 1:251–263, 1973.
40. E. Hüllermeier, S. Destercke, and I. Couso. Learning from imprecise data: Adjustments of optimistic and pessimistic variants. In N. Ben Amor, B. Quost, and M. Theobald, editors, *Scalable Uncertainty Management*, pages 266–279. Springer, 2019.
41. G. Karagiannis. Introduction to Bayesian statistical inference. In L. Aslett, F. Coolen, and J. De Bock, editors, *Uncertainty in Engineering – Introduction to Methods and Applications*, pages 1–14. Springer, 2022.
42. E. Kofler and G. Menges. *Entscheiden bei unvollständiger Information*. Springer, Berlin, 1976.
43. D. Kreiss and T. Augustin. Undecided voters as set-valued information – towards forecasts under epistemic imprecision. In K. Tabia and J. Davis, editors, *Scalable Uncertainty Management*. Springer LNCS, 242–250, 2020.

44. D. Kreiss, M. Nalenz, and T. Augustin. Undecided voters as set-valued information - machine learning approaches under complex uncertainty. In Destercke S. and E. Hüllermeier, editors, *ECML/PKDD 2020 Workshop on Uncertainty in Machine Learning*, 2020. https://sites.google.com/view/wuml-2020/program.

45. I. Levi. *The Enterprise of Knowledge. An Essay on Knowledge, Credal Probability, and Chance*. MIT Press, Cambridge, 1980.

46. R. Little and D. Rubin. *Statistical Analysis with Missing Data*. Wiley, 2nd edition, 2014.

47. C. Manski. Credible interval estimates for official statistics with survey nonresponse. *Journal of Econometrics*, 191:293–301, 2015.

48. Charles Manski. *Partial identification of probability distributions*. Springer, 2003.

49. E. Miranda and S. Destercke. Extreme points of the credal sets generated by comparative probabilities. *Journal of Mathematical Psychology*, 64:44–57, 2015.

50. I. Molchanov and F. Molinari. *Random Sets in Econometrics*. Econometric Society Monographs. Cambridge University Press, 2018.

51. H. Nguyen, V. Kreinovich, B. Wu, and G. Xiang. *Computing Statistics under Interval and Fuzzy Uncertainty: Applications to Computer Science and Engineering*. Springer, Berlin, 2011.

52. M. Oberguggenberger, J. King, and B. Schmelzer. Classical and imprecise probability methods for sensitivity analysis in engineering: A case study. *International Journal of Approximate Reasoning*, 50:680–693, 2009.

53. J. Plass. *Statistical modelling of categorical data under ontic and epistemic imprecision*. PhD thesis, LMU Munich, 2018.

54. J. Plass, M. Cattaneo, T. Augustin, G. Schollmeyer, and C. Heumann. Reliable inference in categorical regression analysis for non-randomly coarsened observations. *International Statistical Review*, 87:580–603, 2019.

55. J. Plass, P. Fink, N. Schöning, and T. Augustin. Statistical modelling in surveys without neglecting 'The undecided'. In T. Augustin, S. Doria, M. Miranda, and E. Quaeghebeur, editors, *ISIPTA '15*, pages 257–266. SIPTA, 2015.

56. U. Pötter. *Statistical Models of Incomplete Data and Their Use in Social Sciences*. Habilitation Thesis, Ruhr-Universität Bochum, 2008.

57. E. Quaeghebeur. Introduction to the theory of imprecise probability. In L. Aslett, F. Coolen, and J. De Bock, editors, *Uncertainty in Engineering – Introduction to Methods and Applications*, pages 37–50. Springer, 2022.

58. F. Ruggeri, D. Ríos Insua, and J. Martín. Robust Bayesian analysis. In D. Dey and C. Rao, editors, *Handbook of Statistics. Bayesian Thinking: Modeling and Computation*, volume 25, pages 623–667. Elsevier, 2005.

59. S. Bi and M. Beer. Overview of stochastic model updating in aerospace application under uncertainty treatment. In L. Aslett, F. Coolen, and J. De Bock, editors, *Uncertainty in Engineering – Introduction to Methods and Applications*, pages 115–130. Springer, 2022.

60. G. Schollmeyer and T. Augustin. Statistical modeling under partial identification: Distinguishing three types of identification regions in regression analysis with interval data. *International Journal of Approximate Reasoning*, 56, Part B:224–248, 2015.

61. N. Shyamalkumar. Likelihood robustness. In D. Rios Insua and F. Ruggeri, editors, *Robust Bayesian Analysis*, pages 127–143. Springer, New York, 2000.

62. E. Tamer. Partial identification in econometrics. *Annual Review of Economics*, 2:167–195, 2010.

63. N. T'Joens, J. De Bock, and G. de Cooman. In search of a global belief model for discrete-time uncertain processes. In J. De Bock, C. de Campos, G. de Cooman, E. Quaeghebeur, and G. Wheeler, editors, *ISIPTA '19*, volume 103 of *Proceedings of Machine Learning Research*, pages 377–385, Thagaste, Ghent, Belgium, 2019. PMLR.

64. M. Troffaes. Imprecise Monte Carlo simulation and iterative importance sampling for the estimation of lower previsions. *International Journal of Approximate Reasoning*, 101:31–48, 2018.

65. S. Vansteelandt, E. Goetghebeur, M. Kenward, and G. Molenberghs. Ignorance and uncertainty regions as inferential tools in a sensitivity analysis. *Statistica Sinica*, 16:953–979, 2006.

66. P. Walley. *Statistical Reasoning with Imprecise Probabilities*. Chapman & Hall, London, 1991.
67. P. Walley. Inferences from multinomial data: Learning about a bag of marbles. *Journal of the Royal Statistical Society. Series B (Methodological)*, 58:3–57, 1996.
68. P. Walley. A bounded derivative model for prior ignorance about a real-valued parameter. *Scandinavian Journal of Statistics*, 24:463–483, 1997.
69. P. Walley. *BI Statistical Methods: Volume I*. Lightning Source, 2015.
70. G. Walter, L. Aslett, and F. Coolen. Bayesian nonparametric system reliability using sets of priors. *International Journal of Approximate Reasoning*, 80:67–88, 2017.
71. G. Walter and T. Augustin. Imprecision and prior-data conflict in generalized Bayesian inference. *Journal of Statistical Theory and Practice*, 3:255–271, 2009.
72. K. Weichselberger. *Elementare Grundbegriffe einer allgemeineren Wahrscheinlichkeitsrechnung I: Intervallwahrscheinlichkeit als umfassendes Konzept*. Physica, 2001.
73. K. Weichselberger. The theory of interval probability as a unifying model for uncertainty. *International Journal of Approximate Reasoning*, 24:149–170, 2000.
74. K. Weichselberger. *Elementare Grundbegriffe einer allgemeineren Wahrscheinlichkeitsrechnung II. Symmetrische Wahrscheinlichkeitstheorie*. (Manuscript), 2013.
75. M. Zaffalon and E. Miranda. Conservative inference rule for uncertain reasoning under incompleteness. *Journal of Artificial Intelligence Research*, 34:757–821, 2009.

Chapter 6
Reliability

Lisa Jackson and Frank P. A. Coolen

Abstract This chapter introduces key concepts for quantification of system reliability. In addition, basics of statistical inference for reliability data are explained, in particular, the derivation of the likelihood function.

6.1 Introduction

Reliability of systems is of crucial importance in all aspects of human life. Systems are understood to be groupings of components in specific structure, with the system functioning depending on the functioning of the components and the system structure. Uncertainty about the functioning of components leads to uncertainty about the system reliability. To study how system reliability depends on the reliability of components, several leading methods from the engineering literature are briefly introduced in Sect. 6.2. In situations of uncertainty about reliability in engineering, appropriate statistical methods are required to deal with the specific nature of data. This chapter provides an overview of such basic methods. Section 6.3 introduces the key statistical concepts, basic statistical models are presented in Sect. 6.4. Throughout the emphasis is on explanation of the likelihood function, which is at the heart of most statistical inference approaches as commonly used in reliability applications. Unknown model parameters can conveniently be estimated by maximisation of the likelihood function, with corresponding theory to assess the uncertainty of the estimates. Bayesian methods for statistical inference are based on the likelihood function as well, hence understanding of the likelihood function is crucial for study of system reliability under uncertainty. Stochastic process models are also crucial to

L. Jackson (✉)
Aeronautical and Automotive Engineering, Loughborough University, Loughborough, United Kingdom
e-mail: l.m.jackson@lboro.ac.uk

F. P. A. Coolen (✉)
Department of Mathematical Sciences, Durham University, Durham, United Kingdom
e-mail: frank.coolen@durham.ac.uk

© The Author(s) 2022
L. Aslett et al. (eds.), *Uncertainty in Engineering*,
SpringerBriefs in Statistics,
https://doi.org/10.1007/978-3-030-83640-5_6

81

describe variable reliability over time, for example, to reflect the effects of maintenance on a system's reliability. A short introduction to such models is presented in Sect. 6.5, again with an emphasis on deriving the likelihood function to enable statistical inference. As this chapter brings together generic introductory material, no specific references are included throughout the text. Instead, the chapter is ended with a brief list of useful resources, pointing to a number of important books which are highly recommended for further reading, and some brief comments about journals in the field.

6.2 System Reliability Methods

When modelling systems there are a number of tools, ranging from combinatorial methods including reliability block diagrams, fault tree analysis and event tree analysis, to more complex methods that cater for a greater range of system characteristics including dependencies, e.g. Markov or Monte Carlo simulation methods. Note when modelling systems where repair times are involved, these typically do not follow the Exponential distribution, e.g. the Lognormal or Weibull distributions may be more suitable. Also including maintenance teams can mean dependencies are introduced via prioritising strategies. In such instances, more complex methods are required. A brief introduction to some of these modelling methods is provided in this section.

6.2.1 Fault Tree Analysis

One of the most common quantification techniques is fault tree analysis. This method provides a diagrammatic description of the various causes of a specified system failure in terms of the failures of its components. The choice of the system failure mode often follows from a failure mode and effects analysis (FMEA). There is the assumption of independence of failures of the components. Logical gates are used to link together events (intermediate events shown as rectangles and basic events representing failure events as circles), where the more common gates include AND or OR, shown in Fig. 6.1, with an example tree shown in Fig. 6.2. Evaluation of the tree using Boolean algebra yields minimal cut sets, denoted as C_i, which are failure combinations of components that are necessary and sufficient to cause failure of the system. Application of kinetic tree theory and the inclusion-exclusion principle (Eq. 6.1) enables the system unavailability (Q_{sys}) performance measure to be calculated, where $P(C_i)$ is the probability of failure of minimal cut set C_i.

Symbol	Name	Causal Relation
	OR	Output event occurs if at least one of the input events occur.
	AND	Output event occurs if all input events occur.
	Vote	Output event occurs if at least m of the input events occur.

Fig. 6.1 Common fault tree gate types

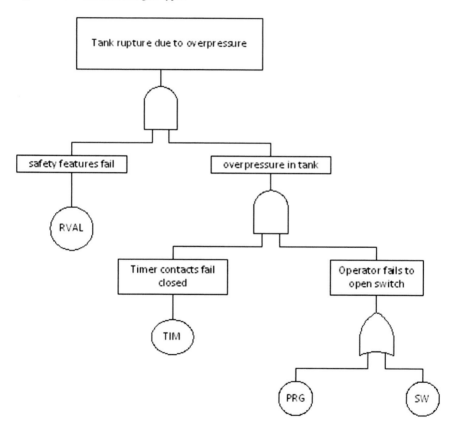

Fig. 6.2 Example fault tree structure

$$Q_{sys} = \sum_{i=1}^{N_c} P(C_i) - \sum_{i=2}^{N_c} \sum_{j=1}^{i-1} P(C_i \cap C_j) + \sum_{i=3}^{N_c} \sum_{j=2}^{i-1} \sum_{k=1}^{j-1} P(C_i \cap C_j \cap C_k) - \dots$$
$$+(-1)^{N_c+1} P(C_1 \cap C_2 \cap \dots \cap C_{N_c}). \tag{6.1}$$

If the unavailability of the system does not meet performance acceptability criteria, then the system must be redesigned. An indication of where to make changes in the system can be achieved by generating component importance measures. This is a measure of the contribution that each component makes to the system failure. One such measure is the Fussel–Vesely measure of importance (I_{FV_i}), defined as the probability of the union of the min cut sets containing each component given that the system has failed, as shown in Eq. 6.2.

$$I_{FV_i} = \frac{P\left(\bigcup\{C_j | i \in C_j\}\right)}{Q_{sys}} \tag{6.2}$$

6.2.2 Fault Tree Extensions: Common Cause Failures

Typically, fault trees have only a limited capability to cater for dependencies within systems. One example is that of common cause failures. Safety systems, for example, often feature redundancy, incorporated such that they provide a high likelihood of protection. However, redundant sub-systems or components may not always fail independently. A single common cause can affect all redundant channels at the same time, examples include ageing (all made at the same time from the same materials), system environment (e.g. pressure or stress related) and personnel (e.g. maintenance incorrectly carried out by the same person). There are several methods to analyse such occurrences, including beta factor, limiting factor, boundary method and alpha method. The beta factor method assumes that the common cause effects can be represented as a proportion of the failure probability of a single channel of the multi-redundant channel. Hence, it assumes that the total failure probability (Q_T) of each component is divided into two contributions: (i) the probability of independent failure, Q_I, and, (ii) the probability of common cause failures, Q_{CCF}. The parameter β is defined as the ratio of the common cause failure probability to the total failure probability, as shown in Eq. 6.3.

$$\beta = \frac{Q_{CCF}}{Q_{CCF} + Q_I} = \frac{Q_{CCF}}{Q_T}. \tag{6.3}$$

There are further extensions to traditional fault tree analysis to cater for a greater range of dependencies, e.g. dynamic fault trees. As the nature of the dependency becomes more complex, other modelling methods will be more suitable. Other types of dependency include standby redundancy, where the probability of failure of the redundant component may change when it starts to function and experience load, hence it is dependent on the operating component and hence failure of both is not statistically independent. Another form is multiple state component failure modes, where a component can exist in more than one state, hence being mutually exclusive cannot be considered in the fault tree (not failing in the open mode does not mean that it works successfully). In such instances, using Markov modelling methods may be desirable. Given the state-space explosion that can exist when using Markov approaches, if there are just small subsections of the system exhibiting the dependency it may be possible to analyse these subsections with Markov methods and use the result embedded in the fault tree approach.

6.2.3 Phased Mission Analysis

As systems become more complex, they can be required to undertake a number of tasks, typically in sequence. An example might include an aircraft flight, where it is required to taxi from the stand, take off, climb to the required altitude, cruise, descend, land and taxi to new destination stand. The collection of tasks can be referred to as a mission, where each task is denoted by a phase which has an associated time period. Mission success requires successful completion of all phases and in addition there may be different consequences resulting from the failure in each phase. For each phase, an appropriate modelling technique can be employed to assess its reliability or availability. When the systems phases are non-repairable then fault tree analysis can be used to assess mission and phase success. For non-repairable scenarios, Markov or Petri nets can be used. The parameter of interest is the mission reliability. It is not appropriate to analyse the reliability of each phase and multiply these together to get the mission reliability because (i) the phases are not independent (i.e. failure in one phase may influence failure in another phase); (ii) the assumption that all components are working at the start of the phase is not correct and (iii) the system can fail on a phase change. For the non-repairable case, the initial step is to construct a mission fault tree. The general form is as shown in Fig. 6.3. The top event is a mutually exclusive OR gate, indicating that only one of the input events must happen to cause the output event (mission failure). When considering failures in phase 2 onwards, you need to include in the fault tree that the mission has been successful up to this point, namely, that it has functioned in the preceding phases. For this reason, you can see the introduction of the NOT gate, i.e. under the intermediate event 'Functions in Phase 1'. Alongside this the failure of the component in earlier phases also needs to be taken into account, hence the component failure is represented as shown in Fig. 6.4. To perform the analysis, both qualitatively and quantitatively, new algebra is required as shown in Fig. 6.5. C_j corresponds to the failure of component C in phase j, where the bar above C_j corresponds to the working state. Considering the success of previous phases $i = 1, \ldots, j - 1$, for failure in phase j, makes the analysis non-coherent, yielding prime implicant sets from a qualitative analysis (necessary and sufficient combinations of events (success and failure)). The size of the problem to be solved can

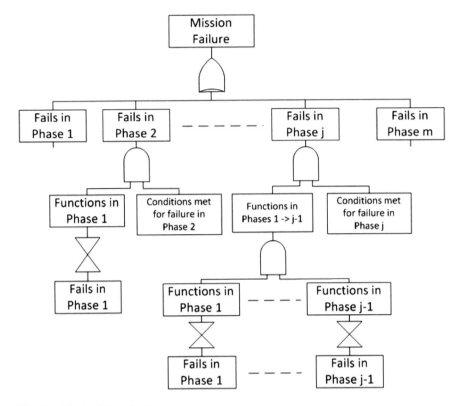

Fig. 6.3 Mission failure fault tree

Fig. 6.4 Revised component
representation

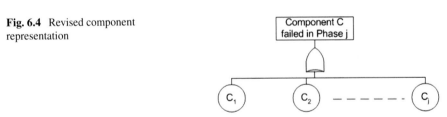

(and usually does) become prohibitively expensive, so approximations are required; these are usually based on coherent approximations of the non-coherent phases (i.e. conversion of prime implicants to minimal cut sets). Approximate quantification formulae (e.g. Rare Event, Minimal Cut Set Upper Bound) can then be used.

When analysing repairable systems, there are two requirements for mission success: (1) the system must satisfy the success requirements throughout each phase period and (2) at the phase change times the system must occupy a state which is successful for both phases involved. The second point implies that we will consider failures on phase transition when calculating mission reliability. Analysis with repairable systems is very similar in terms of generating the mission model and phase

$$\overline{C_1}.\overline{C_2}....\overline{C_{j-1}}.C_j = C_j$$
$$\overline{C_1}.C_1 = 0$$
$$C_i.\overline{C_j} = 0 \quad \text{if} \quad i < j$$

Fig. 6.5 Additional algebra

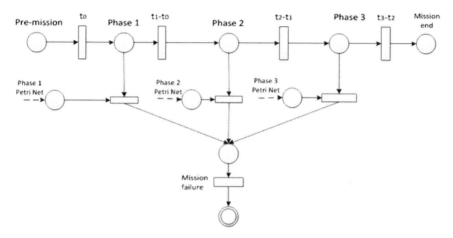

Fig. 6.6 Petri Net representation of a phased mission with three phases

models. This can be achieved using Markov or Petri Net methods. An illustration of a Petri Net example is shown in Fig. 6.6, where the circles represent places and the rectangular boxes represent transitions. Mission and phase reliabilities can be obtained from analysis of the model.

6.3 Basic Statistical Concepts and Methods for Reliability Data

Consider a random quantity $T > 0$, often referred to as a 'failure time' in reliability theory, but it can denote any 'time to event'. Relevant notation for characteristics of its probability distribution includes the cumulative distribution function (CDF) $F(t) = P(T \leq t)$, the survival function $S(t) = P(T > t) = 1 - F(t)$ (also called the reliability function and denoted by $R(t)$), the probability density function (PDF) $f(t) = F'(t) = -S'(t)$ and the hazard rate $h(t) = f(t)/S(t)$. The hazard rate has a possible interpretation with conditioning on surviving time t, for small $\delta t > 0$, $h(t)\delta t \sim P(T \leq t + \delta t \mid T > t)$. Harder to interpret, but also of use, is the cumulative hazard function (CHF) $H(t) = \int_0^t h(x)dx$. Assuming $R(0) = 1$, we get $H(t) = \int_0^t \frac{f(x)}{S(x)}dx = -\ln S(t)$, so $S(t) = \exp\{-H(t)\} = \exp\{-\int_0^t h(x)dx\}$.

A constant hazard rate, $h(t) = \lambda > 0$ for all $t > 0$, gives $S(t) = e^{-\lambda t}$, the Exponential distribution. This can be interpreted as modelling 'no ageing', that is, if an item functions, then its remaining time until failure is independent of its age. This property is unique to the Exponential distribution. An increasing hazard rate models 'wear-out', roughly speaking this implies that an older unit has shorter expected residual life, and decreasing hazard rate models 'wear-in', implying that an older unit has greater expected residual life. It is often suggested that 'wear-out' is appropriate to model time to failure of many mechanical units, whereas electronic units' times to failures may be modelled by 'wear-in'. In a human-life analogy, we can perhaps think about 'wear-in' as modelling time to death at very young age ('infant mortality') and 'wear-out' as modelling time to death at older age, with a period in between where death is mostly 'really random', e.g. caused by accidents. This 'human-life analogy' should only be used for general insight, and is included here as engineers often claim that 'typical hazard rates' for components over their entire possible lifetime are decreasing early on, then remain about constant for a reasonable period, and then become increasing ('bath-tub shaped').

A popular parametric probability distribution for T is defined by the hazard rate $h(t) = \alpha\beta(\alpha t)^{\beta-1}$, for $\alpha, \beta > 0$. This leads to $S(t) = \exp\{-(\alpha t)^{\beta}\}$, and is called a Weibull distribution with scale parameter α and shape parameter β. This distribution is often used in reliability, due to the simple form for the hazard rate. For example, with $\beta = 2$ it models 'linear wear-out' ('twice as old, twice as bad').

An interesting aspect of reliability data is that these are often affected by *censoring*, in particular, so-called *right-censoring*. This means that, instead of actually observing a time at which a failure occurs, the information in the data is a survival of a certain period of time without failing. Clearly, such information must be taken into account, as neglecting it would lead to underestimation of expected failure times.

Two main statistical methodologies use the *likelihood function*, namely, Bayesian methods and maximum likelihood estimation. Hence, derivation of the likelihood function is an important topic in reliability inference. Let $t_1, \ldots, t_n$ be observed failure times, and $c_1, \ldots, c_m$ right-censored observations. For inference on a parameter θ of an assumed parametric model, the likelihood function based on these data is $L(\theta|t_1, \ldots, t_n; c_1, \ldots, c_m) = \prod_{j=1}^{n} f(t_j|\theta) \prod_{i=1}^{m} S(c_i|\theta)$. This actually requires the assumption that the censoring mechanism is independent of the data distribution, if that is not the case the dependence would need to be modelled. It is also possible to consider the likelihood over all possible probability distributions, so not restricting to a chosen parametric model. In this case, the maximum likelihood estimator is the so-called Product-Limit (PL) estimator, presented by Kaplan and Meier in 1958. The theory of counting processes also provides a powerful framework for nonparametric analysis of failure time data, based on stochastic processes and martingale theory. A well-known result within this theory is the Nelson–Aalen estimator for the CHF, which can be regarded as an alternative to the PL estimator.

6.4 Statistical Models for Reliability Data

This is a very wide topic, we can only mention a few important models. We first consider regression models for reliability data. Regression models are generally popular in statistics, and also very useful in reliability applications, where often Weibull models are used, with the survival function depending on a vector of covariates x, and given by $S(t; x) = \exp\left\{-\left(\frac{t}{\alpha_x}\right)^{\eta_x}\right\}$. Some simple forms are often used for the shape and scale parameters as functions of x, e.g. the loglinear model for α_x, specified via $\ln \alpha_x = x^T \beta$, with β a vector of parameters, and similar models for η_x. The statistical methodology is then pretty similar to general regression methods, and implemented in statistical software packages. Such models need to be fully specified, so are less flexible than nonparametric methods, but they allow information in the form of covariates to be taken into account.

Semi-parametric models enable covariates to be taken into account, but do so without fully specifying a parametric model, keeping some more flexibility. Usually, a parametric form for the effect of the covariates on a nonparametric 'baseline model' is assumed. Most famous are the Proportional Hazards (PH) models, presented by Cox in 1972. Here, the hazard rate for covariates x is defined by $h(t; x) = h_0(t)\psi_x$, with $h_0(t)$ the baseline hazard rate (normally left unspecified, so nonparametric), and ψ_x some positive function of x, independent of time t (normally a fully parametric form is assumed for ψ_x). The name of such models results from the fact that $\frac{h(t;x_1)}{h(t;x_2)} = \frac{\psi_{x_1}}{\psi_{x_2}}$, independent of t, so the hazard rates corresponding to different covariates are in constant proportion. For these models, $\ln S(t; x) = -\int_0^t h(u; x)du = -\psi_x \int_0^t h_0(u)du = \psi_x \ln S_0(t)$, so $S(t; x) = [S_0(t)]^{\psi_x}$. These models are used most for survival data in medical applications, but are also common and useful in reliability. As there are no assumptions on the form of the baseline hazard rate, they provide a valuable method to compare the effect of the covariates. An often used PH model is the linear PH model, with $\psi_x = \exp\{x^T \beta\}$, with β a vector of parameters. We now describe the analysis of this particular model, with no further assumptions on $h_0(t)$. The goal is to estimate β and $R_0(t)$, the baseline survival function related to $h_0(t)$. This is far from trivial, as it is not clear how the likelihood function can be derived, since this is neither uniquely defined by a fully specified parametric model, nor completely free as was the case for fully nonparametric models (leading to the PL estimate). Hence, we need to use a different concept.

Suppose we have data on n items, consisting of r distinct event times, $t_{(1)} < t_{(2)} < \ldots < t_{(r)}$ (the case of ties among the event times is a bit more complicated), and $n - r$ censoring times. Let R_i be the risk set at $t_{(i)}$, so all items known to be still functioning just prior to $t_{(i)}$. We can now estimate β via maximisation of the 'likelihood function':

$$L(\beta) = \prod_{i=1}^{r} \frac{\exp\left\{x_{(i)}^T \beta\right\}}{\sum_{l \in R_i} \exp\left\{x_l^T \beta\right\}} \tag{6.4}$$

with $x_{(i)}$ the vector of covariates associated to the item observed to fail at $t_{(i)}$, etc. There have been many justifications for $L(\beta)$, the nicest is the original one by Cox, which is as follows. Let us consider R_i at $t_{(i)}$. The conditional probability that the item corresponding to $x_{(i)}$ is the one to fail at the time $t_{(i)}$, given that there is a failure at $t_{(i)}$, is equal to

$$\frac{h(t_{(i)}; x_{(i)})}{\sum_{l \in R_i} h(t_{(i)}; x_l)} = \frac{\exp\left\{x_{(i)}^T \beta\right\}}{\sum_{l \in R_i} \exp\left\{x_l^T \beta\right\}}.$$

Now $L(\beta)$ is formed by taking the product of all these terms over all failure times, giving a 'likelihood' which is conditional on the event times, sometimes called the 'conditional likelihood' (aka 'partial likelihood' or 'marginal likelihood'). Note that the actual event times $t_{(i)}$ are not used in Eq. 6.4, just the ordering related to the values of the covariates. This relates to the fact that we do not have any knowledge or assumptions about $h_0(t)$. Large sample theory is available for $L(\beta)$, allowing estimation and hypothesis testing similarly as for standard maximum likelihood methods.

Next, one must consider estimation of the survival function. Once the estimate for β has been derived, let us denote this by $\hat{\beta}$, it is possible to obtain a nonparametric estimate of the baseline survival function. Let

$$\hat{S}_0(t) = \prod_{j:t_{(j)} \leq t} \hat{\alpha}_j,$$

where the $\hat{\alpha}_j$'s are derived via

$$\alpha_j^{\lambda_j} = 1 - \frac{\lambda_j}{\sum_{l \in R_j} \lambda_l},$$

and

$$\lambda_j = \exp\left\{x_j^T \beta\right\},$$

and taking $\beta = \hat{\beta}$. This actually gives the maximum likelihood estimate for the survival function, under the assumption that β is indeed the given estimate $\hat{\beta}$.

6.5 Stochastic Processes in Reliability—Models and Inference

Suppose we have a system which fails at certain times, where there may be some actions during this period which affect failure behaviour, e.g. minimal repairs to allow the system to continue its function, or replacement of some components, or other improvements of the system. Let the random quantities $T_1 < T_2 < T_3 < \ldots$

be the times of failure of the system, and let $X_i = T_i - T_{i-1}$ (with $T_0 = 0$) be the time between failures $i - 1$ and i. These X_i, or in particular trends in these, are often of main interest in analysis of system failure, e.g. to discover whether or not a system is getting more or less prone to fail over time. Hence, the major concern is often detection and estimation of trends in the X_i. Therefore, we cannot just assume these X_i's to be (conditionally) independent and identically distributed (iid), as is often assumed for standard statistical inference on such random quantities. Instead, we need to consider the process in more detail. A suitable characteristic for such a process is the so-called 'rate of occurrence of failure' (ROCOF). Let $N(t)$ be the number of failures in the period $(0, t]$, then the ROCOF is

$$v(t) = \frac{d}{dt} EN(t).$$

An increasing (decreasing) ROCOF models a system that gets worse (better) over time. Of course, all sorts of combinations can also be modelled, e.g. first a period of decreasing ROCOF, followed by increasing ROCOF, to model early failures after which a system improves, followed by a period in which the system wears out. Note that the ROCOF is not the same as the hazard rate (the definitions are clearly different!), although intuitively they might be similar. If we consider a standard Poisson process, with iid times between failures being Exponentially distributed, then the ROCOF and hazard rate happen to be identical.

An estimator for $v(t)$ is derived by defining a partition of the time period of interest, counting the number of failures in each of the intervals of this partition, and dividing this number by the length of the corresponding interval if necessary (i.e. if not all intervals are of equal length). However, it is more appealing to use likelihood theory for statistical inference, which we explain next for nonhomogeneous Poisson processes (NHPP), for which the ROCOF is a central characteristic often used explicitly to define such processes.

NHPP are relatively simple models that can be used to model many reliability scenarios, and for which likelihood-based statistical methodology is well developed and easy to apply. The crucial assumption in these models is that the numbers of failures in distinct time intervals are independent if the process characteristics are known. A NHPP with ROCOF $v(t)$ is easiest defined by the property that the number of failures in interval $(t_1, t_2]$ is a Poisson distributed random quantity, with mean

$$m(t_1, t_2) = \int_{t_1}^{t_2} v(t) dt.$$

This implies that the probability of 0 failures in interval $(t_1, t_2]$ equals $\exp\{-m(t_1, t_2)\}$, and the probability of 1 failure in this interval equals $m(t_1, t_2) \exp\{-m(t_1, t_2)\}$. Of course, if $v(t)$ is constant we have the standard Poisson process. For statistical inference, we wish to find the likelihood function corresponding to a NHPP model, given failure data of a system. Suppose we have observed the system over time period $[0, r]$, and have observed failures at times $t_1 < t_2 < \ldots < t_n \leq r$, assuming there were no tied observations (so only a single failure at each failure time; else things get slightly more complicated). The likelihood function is derived in a similar way as for iid data, that is, the reasoning of using PDFs at failure times in the likelihood for such data. Let $\delta t_i > 0$, for $i = 1, \ldots, n$, be very small. The process observed can then be described as consisting of: 0 failures in $(0, t_1)$, and 1 failure in $[t_1, t_1 + \delta t_1)$, and 0 failures in $[t_1 + \delta t_1, t_2)$, etc., until no failures in $[t_n + \delta t_n, r]$ (this last bit is just deleted if $r = t_n$, so when observation of the process is ended at the moment of the n-th failure). To derive the corresponding likelihood function, we take the product of the probabilities for these individual events, so

$$\exp\{-m(0, t_1)\} \times m(t_1, t_1 + \delta t_1) \exp\{-m(t_1, t_1 + \delta t_1)\} \times \exp\{-m(t_1 + \delta t_1, t_2)\} \times \ldots$$
$$\ldots \times \exp\{-m(t_n + \delta t_n, r)\}$$
$$= \left\{ \prod_{i=1}^{n} \left[\int_{t_i}^{t_i + \delta t_i} v(t)dt \right] \right\} \times \exp\left[-\int_0^r v(t)dt \right].$$

Now, use that for very small δt_i, we have that

$$\int_{t_i}^{t_i + \delta t_i} v(t)dt \approx v(t_i)\delta t_i.$$

Now we divide through by $\prod_{i=1}^{n} \delta t_i$, and let all $\delta t_i \downarrow 0$ (this is, exactly the same that leads to the PDFs appearing in the likelihood function for iid data). This gives the likelihood function, for this model and based on these n failure data and observation over the period $[0, r]$:

$$L = \left\{ \prod_{i=1}^{n} v(t_i) \right\} \exp\left[-\int_0^r v(t)dt \right].$$

For optimisation, it is easier to use the log-likelihood function, which is also needed for related statistical inference, and which is equal to:

$$l = \sum_{i=1}^{n} \ln v(t_i) - \int_0^r v(t)dt.$$

It is possible to work with this likelihood non-parametrically, but often one assumes a parametric form for the ROCOF, making maximum likelihood estimation again conceptually straightforward (although it normally requires numerical optimisation). Two simple, often used parametric ROCOFs are

$$v_1(t) = \exp(\beta_0 + \beta_1 t)$$

and

$$v_2(t) = \gamma \eta t^{\eta-1},$$

with $\gamma, \eta > 0$.

Many models that have been suggested, during about the last three decades, for software reliability, are NHPPs which model the software testing process as a fault counting process. A famous model was proposed by Jelinski and Moranda in 1972, which is based on the following assumptions: (1) software contains an unknown number of bugs, N; (2) at each failure, one bug is detected and corrected; (3) the ROCOF is proportional to the number of bugs present. So, they use a NHPP with failure times $T_i, i = 1, \ldots, N$ and $T_0 = 0$, defined by

$$v(t) = (N - i + 1)\lambda, \quad \text{for } t \in [T_{i-1}, T_i),$$

for some constant λ. Then N and λ are both considered unknown, and estimated from data, where, of course, inference for N tends to be of most interest, or, in particular, the number of remaining bugs. Many authors have contributed to such theory by changing some model assumptions. For example, non-perfect repair of bugs has been considered, and even the possibility of such repair introducing new bugs (possibly a random number), for this last situation so-called 'birth-death processes' can be used. Also non-constant λ has been considered, e.g. with the idea that some bugs may tend to show earlier than others. Also Bayesian methods for such models, and even software reliability models more naturally embedded in Bayesian theory, have been suggested and studied. However, although there is an enormous amount of literature in this area, as indeed mathematical opportunities appear to have no limit here, the practical relevance of such models seems to be rather limited and few interesting applications of such models have been reported in software reliability. Recently, the important topic of testing of reliability of systems including software has received increasing attention, which is much needed to ensure reliable systems.

Useful sources

P.K. Andersen, O. Borgan, R.D. Gill and N. Keiding, *Statistical Models Based on Counting Processes* (Springer, 1993).
T. Aven and U. Jensen, *Statistical Models in Reliability* (Springer, 1999).
R.E. Barlow, *Engineering Reliability* (SIAM, 1998).
R.E. Barlow and F. Proschan, *Mathematical Theory of Reliability* (Wiley, 1965).
T. Bedford and R. Cooke, *Probabilistic Risk Analysis: Foundations and Methods* (Cambridge University Press, 2001).
P. Hougaard, *Analysis of Multivariate Survival Data* (Springer, 2000).
R.S. Kenett, F. Ruggeri and F.W. Faltin (Eds), *Analytic Methods in Systems and Software Testing* (Wiley, 2018).
J.F. Lawless, *Statistical Models and Methods for Lifetime Data* (Wiley, 1982).
R.D. Leitch, *Reliability Analysis for Engineers* (Oxford University Press, 1995).
H.F. Martz and R.A. Waller, *Bayesian Reliability Analysis* (Wiley, 1982).

W.Q. Meeker and L.A. Escobar, *Statistical Methods for Reliability Data* (Wiley, 1998).
N.D. Singpurwalla and S.P. Wilson, *Statistical Methods in Software Engineering: Reliability and Risk* (Springer, 1999).

Leading international journals in this field include Reliability Engineering and System Safety, IEEE Transactions on Reliability, Journal of Risk and Reliability, Quality and Reliability Engineering International. Statistical methods for reliability data are presented in a wide variety of theoretical and applied Statistics journals, theory and methods for decision support are often published in the Operations Research literature.

Chapter 7
Simulation Methods for the Analysis of Complex Systems

Hindolo George-Williams, T. V. Santhosh, and Edoardo Patelli

Abstract Everyday systems like communication, transportation, energy and industrial systems are an indispensable part of our daily lives. Several methods have been developed for their reliability assessment—while analytical methods are computationally more efficient and often yield exact solutions, they are unable to account for the structural and functional complexities of these systems. These complexities often require the analyst to make unrealistic assumptions, sometimes at the expense of accuracy. Simulation-based methods, on the other hand, can account for these realistic operational attributes but are computationally intensive and usually system-specific. This chapter introduces two novel simulation methods: **load flow simulation** and **survival signature simulation** which together address the limitations of the existing analytical and simulation methods for the reliability analysis of large systems.

7.1 Introduction

A system is classed as complex from one of two fronts—in terms of the functional relationships between its components and in terms of its structure. A structurally complex system does not conform to a series, parallel, or series-parallel configuration. Most real-world systems are composed of components that can operate at multiple

H. George-Williams
Institute of Energy and Sustainable Development, School of Engineering and Sustainable Development, De Montfort University, Oxford, United Kingdom
e-mail: hindolo.george-williams@dmu.ac.uk

T. V. Santhosh
Institute for Risk and Uncertainty, University of Liverpool, Liverpool, United Kingdom
e-mail: s.santhosh@liverpool.ac.uk

Bhabha Atomic Research Centre, Mumbai, India

E. Patelli (✉)
Centre for Intelligent Infrastructure, Civil and Environmental Engineering, University of Strathclyde, Glasgow, United Kingdom
e-mail: edoardo.patelli@strath.ac.uk

© The Author(s) 2022
L. Aslett et al. (eds.), *Uncertainty in Engineering*,
SpringerBriefs in Statistics,
https://doi.org/10.1007/978-3-030-83640-5_7

95

performance levels or states and components with a functional coupling with other components. Such systems are deemed functionally complex, since their states cannot be directly deduced from their traditional two-state structure functions. They are characterised by multiple states, with the number of states determined by the diversity in the states of their components, structure and the functional relationships between their components [21]. In these systems, the number of performance levels may or may not be finite, depending on the performance measure under consideration and the type of system [21]. For instance, the power generated by a power plant may take any value between zero and its maximum achievable value, depending on the performance levels of its component and the demand on the grid. Complex systems may be standalone or form an indispensable part of some critical system like healthcare, safety-critical and industrial control systems. It is, therefore, important to be able to assess their susceptibility to failures, as well as quantify and predict the ensuing consequences, for effective planning of restoration and mitigation measures.

7.2   Reliability Modelling of Systems and Networks

In system reliability evaluation, the analyst has numerous techniques at their disposal, which can be classified as heuristic-, analytical- or simulation-based [1] and further as static or dynamic. In particular, dynamic techniques not only model the system based on the functional and structural relationships between its components, but also support dynamic relationships like inter-component and inter-system dependencies.

7.2.1   Traditional Approaches

Reliability Block Diagrams and **Fault Trees** have been extensively used in the reliability evaluation of binary-state systems. Both techniques have proven particularly useful for moderately sized systems with series-parallel configurations. However, they become difficult to apply with large or complex systems and often require additional techniques to decompose the system. The Reliability Graph [40] was, therefore, developed to overcome this difficulty and proved very efficient in modelling structural complexities. Reliability block diagrams, fault trees and reliability graphs, however, assume components to be statistically independent, which renders them inadequate for systems susceptible to restrictive maintenance policies and inter-component dependencies. However, techniques including but not limited to dynamic reliability block diagrams [10], dynamic fault trees [6], condition-based fault trees [35], dynamic flow graphs [2], Petri Nets [26] and other combinatorial techniques [38] have been developed to model these dynamic relationships. They have found application in a wide range of reliability engineering problems, including repairable systems with restrictive maintenance policies.

Though the earliest forms of these techniques including binary decision diagrams were applicable only to binary-state systems, numerous instances of their recent extension to multi-state systems exist, see, e.g. [39]. However, these extensions either *require state enumeration or the derivation of the minimal path or cut sets of the system, which is an NP-hard problem* [41].

The **extended block diagram technique** and **graph-based algorithms** share two common limitations. First, they define reliability with respect to the maximum flow through the system. Therefore, they are limited to systems with single output nodes and those with multiple output nodes where only the presence of flow at these nodes is relevant and not the relative magnitude of the flow. The second limitation arises from the assumption that there are no flow losses in the system, making them inapplicable to certain practical systems like energy systems and pipe networks, susceptible to losses in some failure modes. More recently, various researchers have made invaluable contributions to multi-state system reliability analysis, developing techniques applicable to a wide range of systems [22]. These techniques have mainly been based on either the structure function approach, stochastic process, simulation or the Universal Generating Function approach [21, 25].

The most popular stochastic process employed in system reliability analysis is the **Markov Chain (MC)**, which involves enumerating all the possible states of the system and evaluating the associated state probabilities [25]. This technique is only easily applicable to exponential transitions or distributions with simple cumulative distribution functions, requires complicated mathematics and becomes complex for large systems. For an M component binary-state system, the number of states in the model ranges from $M + 1$ for series systems, to 2^M for parallel systems. For large multi-state systems, the number of states increases exponentially, rendering the model difficult to construct and expensive to compute.

The **Universal Generating Function** was introduced to address the state explosion problem of the MC. It allows the algebraic derivation of a system's performance from the performance distribution of its components [21, 24]. However, both the Universal Generating Function and Markov Chain are limited in the number of reliability indices they can quantify. Also, like all multi-state system reliability evaluation techniques, they are maximum-flow-based and assume flow conservation across components. The Universal Generating Function, though straightforward for series/parallel systems, it requires a substantial effort for complex topologies.

Simulation methods are the most suitable for multi-state system reliability and performance evaluation, since they mimic the actual operation of systems. Their advantage over their analytical counterpart is due to the fact that they support any transition distribution, allow the effects of external factors on system performance to be investigated [43] and are easily integrated with other methods [36]. In particular, they allow the explicit consideration of the effects of uncertainty and imprecision on the system, providing a powerful tool for risk analysis and by extension, rational decision-making under uncertainty. They are, therefore, mostly used to analyse systems for which analytical approaches are inadequate. However, even some of the existing simulation methods [23, 43] require prior knowledge of the system's path set, cut set or structure function and are mostly limited to binary-state systems [42].

7.2.2 Interdependencies in Complex Systems

Engineers and system designers are under immense pressure to build systems robust and adequate enough to meet the ever-increasing human demand and expectation. Unavoidably, the resulting systems are complex and highly interconnected, which ironically constitute a threat to their resilience and sustainability [18]. Two systems are *interdependent* if at least a pair of components (one from each system) are coupled by some phenomena, such that a malfunction of one affects the other. In such systems, an undesirable glitch in one system could cascade and cause disruptions in the coupled system. The cascade could be fed back into the initiating system and the overall consequences may be catastrophic [5]. To minimise the effects of failures, some interdependent systems are equipped with *reconfiguration* provisions. This normally entails transferring operation to another component, rerouting flow through alternative paths, or shutting down parts of the system.

The achievement of maximum overall system performance is, in general, desirable. However, in many applications (nuclear power plants, for instance), it is more important to guarantee system availability and recovery in the shortest possible time, following component failure [16]. *Interdependencies* are manifested in engineering systems at two levels: between components (*inter-component*), which can be functional or induced and between systems/subsystems (*inter-system*) [15].

Functional dependencies are due to the topological and/or functional relationships between components. Induced dependencies, on the other hand, are due to a state change in one component (the initiator) triggering a corresponding state change in another (the induced), such that even when the initiator is reinstated, the induced does not reinstate, unless manually made to do so. Functional dependencies in standalone systems are intrinsically accounted for by the innate attributes of the system reliability modelling and evaluation technique while induced dependencies require explicit modelling. Inter-system dependencies, on the other hand, are due to functional or induced couplings between multiple systems. The functional dependencies in these systems, however, may require explicit modelling. This is the case for components relying on material generated by another system. For instance, an electric pump in a water distribution system relies on the availability of the electricity network.

Induced dependencies are further divided into *Common-Cause Failures* (CCF) [27] and cascading events, as summarised in Fig. 7.1. Common-cause failures are the simultaneous failure of multiple similar components due to the same root cause. Their origin is traceable to a coupling that normally is external to the system. Notable instances are shared manufacturing lines, shared maintenance teams, shared environments and human error. A group of components susceptible to the same CCF event is called a Common-Cause Group (CCG). An important point to note about common-cause failures is that, on occurrence of the failure event, there is a probability associated with multiple component failure and that the affected components fail in the same mode. Consequently, the number of components involved in the event ranges from 1 to the total number of components in the CCG. CCF events may affect an entire system or only a few of its components and, therefore, pose a consider-

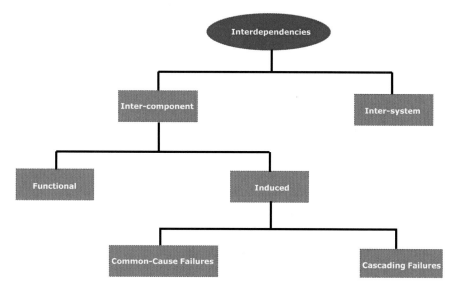

Fig. 7.1 Types of interdependencies in complex systems. Functional dependencies—such as when the failure of power supply forces the unavailability of connected components. Common-Cause Failures—due to earthquake excitation, vibration, environmental conditions (temperature, humidity, contaminants), shared maintenance. Cascading events such as the failure of one component might overload other components

able threat to the reliability of systems. CCF modelling and quantification attracts keen interest from system reliability and safety researchers, as well as practitioners. Examples of the work that has been done in this field can be found in [28, 33, 37]. Most of the methods presented in these publications, however, are built on reliability evaluation techniques that do not segregate the topological from the probabilistic attributes of the system. As such, they are computationally expensive for problems involving multiple reliability analysis of the same system. They also have yet to be applied to multi-state systems, as well as systems susceptible to both cascading and common-cause failures.

Cascading failures are those with the capacity to trigger the instantaneous failure of one or more components of a system. They can originate from a component or from a phenomenon outside the system boundary. The likelihood of the initiating event originating from within the system distinguishes them from CCF. Another point of dichotomy is that the affected components do not necessarily have to be similar or fail in the same mode. In addition, at the occurrence of the initiating event, the probability of all the coupled components failing is unity, same for the case when they are in a state rendering them immune [15, 18]. A few prominent examples of initiating events external to the system are extreme environmental events, natural disasters, external shocks, erroneous human-system interactions and terrorist acts. Various models have been developed to study the effects of cascading failures on complex systems [29]. However, a good number of these models only assess their response to targeted attacks, variation in some coupling factor or the relative importance of system components. When faced with the additional situation of random component failures, a complete reliability and availability analysis should be performed [18]. Even methods that fulfill this requirement have their applicability hampered by components that undergo non-Markovian transitions, components susceptible to delayed transitions, and reconfigurable systems.

7.3   Load Flow Simulation

The load flow simulation is a recently proposed technique for the reliability and performance analysis of multi-state systems [17]. It is based on the fact that if the performance levels of a system's components are known, the performance levels of the system can be directly derived from its network model. In this formalism, each component is modelled as a semi-Markov stochastic process and the system as a directed graph whose nodes are the components of the system. The approach is intuitive and applicable to any system architecture and easily programmable on a computer. It outperforms other multi-state system reliability analysis approaches, since it does not require state enumeration or cut set definition. Efficient algorithms for manipulating the adjacency matrix of this directed graph to obtain the flow equations of the system are available in OpenCossan [31].

The operation of the system is simulated using Kinetic Monte Carlo method by initially sampling the state and time to the next transition (hereafter referred to as transition parameters) of each component. The simulation jumps to the smallest sampled transition time t_{min}, at which time the states of the components undergoing the transition are updated. Using the updated performance levels of the components of the system, the virtual flow across the system is computed via a linear programming procedure that employs the interior-point algorithm. The new transition parameters of the components undergoing a transition are then sampled and the simulation jumps to the next smallest transition time. This cycle of component transition parameter sampling, transition forcing and system performance computing continues until the mission time T is reached. The system performance computed at every component transition is captured and saved in counters, from which the performance indices of the system can be deduced. A component shutdown and restart procedure is incorporated to replicate the actual operating principles of most practical systems. In this procedure, the availability of each system component is tested against its predefined reference minimum input load level at every transition and the effects of functional interdependence on the failure probability of the components are accounted for. Figure 7.2 provides a high-level illustration of the load flow simulation procedure.

Ageing and component performance degradation is common in most systems. For such systems, techniques built around the flow conservation principle become obsolete, as the flow generated by sources can be dissipated in intermediate components in certain failure modes. For instance, consider a 100 MW power generator supplying a 95 MW load through a 125 MW transformer. If there are no power losses

Fig. 7.2 Flowchart of the load flow simulation

in the transformer, 95 MW will be drawn from the generator and delivered to the load. However, if the efficiency of the transformer deteriorates to say 75%, it now takes all 100 MW from the generator but delivers only 75 MW to the load. In both cases, the apparent difference between the generation capacity and demand is the same but the power drawn from the generator increases while the effective power supplied to the load deteriorates. For this example, the demand would have to be slashed to 75 MW or less, to preserve the operational integrity of the generator. Other scenarios where component inefficiency affects system reliability are: a power transmission line prone to losses and an oil pipeline where a failure mode is a hole in a pipe or gasket failure at some flange [17].

The load flow simulation approach has been successfully applied to the availability assessment of a reconfigurable offshore installation [18], dynamic maintenance strategy optimization of power systems [19] and the probabilistic risk assessment of station blackout accidents in nuclear power plants [16].

> **Advantages Over Existing Techniques:**
>
> 1. Inherits all the advantages of simulation approaches used for system reliability and performance evaluation.
> 2. Implements any system structure with relative ease, since it doesn't require knowledge of the minimal path or cut sets prior to system analysis.
> 3. Calculates the actual flow across every node of the system.
> 4. Models systems made up of multiple source and sink nodes with competing static or dynamic demand.
> 5. Models losses in components and across links.
> 6. Models component restart and shutdown.
> 7. Not limited to integer-valued node capacities and system demand, as required by other graph-based algorithms.

7.3.1 Simulation of Interdependent and Reconfigurable Systems

Load flow simulation allows the modelling of inter-component and inter-system dependencies, thereby supporting the reliability assessment of realistic engineering systems [18]. Components and external events that influence the operation of the

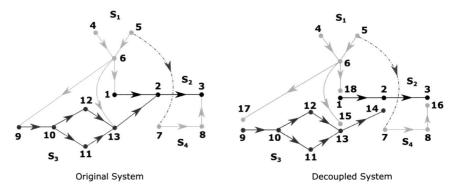

Fig. 7.3 Illustration of decoupling procedure for interdependent systems

system are identified and numbered, followed by the identification and modelling of all the inter-component dependencies in the system. The strategy is to decouple the interdependent system into its constituent systems (subsystems) as shown in [18]. The nodes associated with each subsystem are then identified and its directed graph obtained (i.e. only nodes with actual commodity flow are considered). The states of each node are then identified and modelled as described in [17].

For illustrative purposes, consider the original system in Fig. 7.3 (left panel). It is an interdependent four commodity system—each solid line transports a commodity and the broken line depicts an induced dependency in the direction of the arrow. Node 2 is part of subsystem S_2 and relies the commodity from subsystem S_3 to drive its operation. One would say it is functionally dependent on subsystem S_3 and exhibits a dual operation mode, operating both as a sink and an intermediate node. Its sink mode directly influences flow in S_3, while its transmission mode directly influences flow in S_2. It is, therefore, logical to separate node 2 into its constituent nodes, each representing a mode of operation. **Virtual nodes** representing the *sink modes* of dual nodes are created and assigned new IDs, creating a decoupled system (see Fig. 7.3 (right panel)). A load-source functional dependency exists between the decoupled nodes, since the transmission node is incapacitated if flow into the sink node is inadequate. Therefore, they make a load-source pair, with the transmission node being the load and the sink node, the local source node.

Local sources, otherwise known as *support nodes* in load-source pairs, are modelled as binary-state objects: state 1 (active) has capacity l, depicting the availability of the dependent node; State 2 (inactive) has capacity 0 and depicts its unavailability. l is the minimum level of support required to operate the dependent/sink node and in practical cases represents the load rating of that component. By applying the decoupling procedure described to all load dependency relationships in the system, the following load-source pairs; {2, 14}, {3, 16}, {1, 18}, {13, 15} and {9, 17} are obtained. $\mathbb{L}_i = \{j, l\}$ signifies that node i requires a minimum of l units of a certain commodity from node j to operate. If i has a load dependency relationship with multiple nodes, $\mathbb{L}_i$ takes the form of a two-column matrix, with each row defining the node's relationship with another node.

Induced dependencies are defined by the parameter $D_i = \{d_{j1}, d_{j2}, d_{j3}, d_{j4}\}_{u \times 4} \mid$ $j = 1, 2, ..., u - 1, u$, which defines the state change induced in other nodes as a result of a state change in node i. d_{j1} is the state of i triggering the cascading event, d_{j2}; the affected node, d_{j3}; the state the node has to be in to be affected, and d_{j4}; its target state on occurrence of the event. Each row of D_i defines the behaviour of an affected node, and u, the number of relationships. If node i and the affected node d_{j2} belong to different subsystems, the subsystem the latter belongs to is dependent on the subsystem of the former. For example, suppose state 2 of node 5 in Fig. 7.3 forces node 7 into state 3 if it is in state 1 at the time node 5 makes the transition to state 2. The induced dependency of node 7 on node 5 is defined by D_5 as

$$D_5 = \begin{pmatrix} 2 & 7 & 1 & 3 \end{pmatrix} \tag{7.1}$$

Once the system has been decoupled, the dependency tree depicting the relationships between its subsystems and their ranking is derived. The rank of a subsystem depends on its position on the tree relative to the reference subsystem. The independent subsystem, which is also the reference subsystem, is assigned rank 1 and the remainder ranked in ascending order of their longest distance from this reference. See [18] for the details of the ranking, reconfiguration and simulation procedures.

7.3.2 Maintenance Strategy Optimization

The load flow simulation approach can be exploited to optimise the maintenance strategies of complex systems. The multi-state semi-Markov models of components are extended to represent their behaviour under various maintenance strategies. The operation of the system is then simulated using a slightly modified version of the simulation procedure depicted in Fig. 7.2 and detailed in [19]. Non-Markovian component transitions associated with the operational dynamics imposed by maintenance strategies are implemented. For example, the maintenance of a failed component can only be initiated if there is an idle maintenance team, making the transition of the component from its failed to working state non-Markovian, since it is conditional on the availability of a maintenance team. Additional component states such as preventive maintenance, corrective maintenance, shutdown, diagnostics, idle and awaiting maintenance are included to model different maintenance activities.

To illustrate the derivation of the multi-state model of a component under various maintenance strategies, consider a binary-state component. The component is subject to both preventive and corrective maintenance and maintained by a limited number of maintenance teams. In addition, its corrective maintenance consists of two stages: a diagnosis stage and a restoration stage. Following diagnosis, the maintenance team could proceed with the actual repairs if spares are not required or make a spares request. There is a known probability associated with spares being needed for a repair

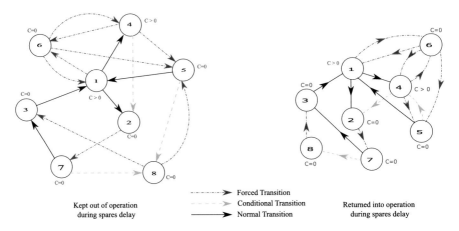

Fig. 7.4 Multi-state models of binary-state component under maintenance delays

and while the maintenance team awaits the spares, it could be assigned to another component. Similarly, there is a probability associated with spares being needed to complete the preventive maintenance of the component, which could be interrupted if these spares are not immediately available. The resulting multi-state models of the component under two contrasting maintenance strategies are shown in Fig. 7.4, with the component's state assignments and possible transitions. Transitions are either normal, forced or conditional. Normal transitions occur randomly and depend only on their associated time-to-occurrence distributions. Forced transitions occur purely as a consequence of events outside the component boundary, and their time-to-occurrence distributions are unknown. Conditional transitions, on the other hand, have a known time-to-occurrence distribution but are assigned a lower priority and only occur on fulfilment of a predefined probabilistic condition or set of conditions [19]. Unlike normal transitions in which the next state of the component depends only on its current state, the next state of the component under forced transitions may also depend on its previous state. As such, the multi-state component transition parameter sampling procedure presented in [17] cannot be used to determine the transition parameters of the component. For this, the set of procedures presented in [19] are required. The binary-state component models in Fig. 7.4 can be generalised for multi-state components by defining one 'Idle' state (if components are kept out of operation during spares delay), a 'Diagnosis' state (where necessary) and one 'Corrective Maintenance' state for each repairable failure mode.

With this approach, multiple contrasting complex maintenance strategies can be simulated without the need to modify the simulation algorithm, as the maintenance strategy is implemented at the component level. See, for instance, the optimal maintenance strategies for a hydroelectric power plant derived in [19].

7.3.3 Case Study: Station Blackout Risk Assessment

The complete lack of AC power at a nuclear power plant is critical to its safety, since AC power is required for its decay heat removal. Though designed to cope with these incidents, nuclear power plants can only do so for a limited time. The impact of station blackouts on a nuclear power plant's safety is determined by their frequency and duration. These quantities, however, are traditionally computed via a static fault tree analysis that deteriorates in applicability with increasing system complexity. The load flow simulation approach was used to quantify the probability and duration of possible station blackouts at the Maanshan Nuclear Power Plant in Taiwan, accounting for interdependencies between system components, maintenance, system reconfiguration, operator response strategies and human errors [16].

The Maanshan Plant is powered through two physically independent safety buses, which themselves are powered by six offsite power sources through two independent switchyards. Each safety bus has a dedicated backup diesel generator and both buses share a third diesel generator. Two gas turbine generators connected through the second switchyard power the plant's safety systems if all three diesel generators are unavailable. The gas turbine generators, however, take about 30 min to become fully operational, when powered on. The goal in this case study was to quantify the risk to the plant, of station blackouts initiated by the failure of the grid sources, as well as the switchyards and identify the best recovery strategy, to minimise this risk.

The load flow simulation approach was used to model the structural/functional relationships between the components of the system as described in Sect. 7.3 and the formalism described in Sect. 7.3.1 to model both the interdependencies between components and their dynamic behaviour under various recovery strategies. The full details of the solution approach and results are available in [16].

7.4 Survival Signature Simulation

For very large-scale systems and networks, the full system structure information (or structure function, minimal paths sets, etc.) might not be available or may be difficult to obtain. Having a compact representation of the system, therefore, is advantageous.

Survival signature [7] has been proposed as a generalisation of system signature [11, 12] to quantify the reliability of complex systems consisting of independent and identically distributed (*iid*) or exchangeable components, with respect to their random failure time. It has been shown in [8] how the survival signature can be derived from the signatures of two subsystems in both series and parallel configuration. The authors developed a non-parametric-predictive inference for system reliability using the survival signature. Aslett et al. [3] demonstrated the applicability of the survival signature to system reliability quantification via a parametric, as well as non-parametric approach. An efficient computational approach for computing approximate and exact system and survival signatures has been recently presented in [20, 34]. Feng et al. [13] developed an analytical method to calculate the sur-

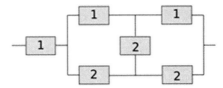

Fig. 7.5 Example of a bridge network composed of six-component of two types

Table 7.1 Survival signature for the system shown in Fig. 7.5

l_1	l_2	$\Phi(l_1, l_2)$	l_1	l_2	$\Phi(l_1, l_2)$
0	[0, 1, 2, 3]	0	2	[0, 1]	0
1	[0, 1]	0	2	2	1/3
1	2	1/9	2	3	2/3
1	3	1/3	3	[0, 1, 2, 3]	1

vival function of systems with uncertainty in the parameters of component failure time distributions. These methods are all useful but less practical for larger complex systems and not applicable to non-exponential transitions.

As an illustration, consider a six-component bridge network with two component types (Fig. 7.5), the survival function is given by Table 7.1.

Considering 2 working components of type 1; $l_1 = 2$ and 3 of type 2; $l_2 = 3$, there are three possible combinations in total but only two combinations lead to success (the survival of the system) of the system. Hence, the survival signature of the system is $\frac{2}{3}$, as shown in Table 7.1. Similarly, for $l_1 = 3$ and $l_2 = [0, 1, 2, 3]$, there are eight possible combinations in total, all of which result in success. Hence, the survival signature of the system in this case is equal to 1.0. Thus, knowing the success paths from the combinations of multiple types of active components, it is possible to compute the survival function of a complex system.

Exact analytical solutions are restricted to particular cases (e.g. systems with component failure times following the exponential distribution and non-repairable components). The survival function of a system with K component types is given by

$$P(T_s > t) = \sum_{l_1=0}^{m_1} \cdots \sum_{l_K=0}^{m_K} \phi(l_1, \ldots, l_K) P(\bigcap_{k=1}^{K} \{C_k(t) = l_k\}) \qquad (7.2)$$

where

$$P(\bigcap_{k=1}^{K} \{C_k(t) = l_k\}) = \prod_{k=1}^{K} \binom{m_k}{l_k} [F_k(t)]^{m_k - l_k} [1 - F_k(t)]^{l_k} \qquad (7.3)$$

Here, $C_k(t) \in \{0, 1, \ldots, m_k\}$ denotes the number of components of type k in the system which function at time t, and $F_k(t)$ represents the CDF of the random failure times of components of the different types. In this approach, we have a strong *iid*

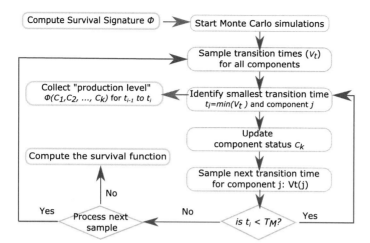

Fig. 7.6 Flow chart of the Monte Carlo simulation algorithm for complex systems with repairable components based on survival signature. Details of the simulation method are available in [30]

assumption of failure times within same components types. With this assumption, all state vectors [7] are equally likely to occur.

However, simulation methods can be applied to study and analyse any system, without introducing simplifications or unjustified assumptions. A Monte Carlo-based approach can be combined with survival signature, to estimate the reliability of a system in a simple and efficient way. A possible system evolution is simulated by generating random events (i.e. the random transition such as failure times of the system components) and then estimating the status of the system based on the survival signature (Eq. (7.2)). By counting the number of occurrences of a specific condition (e.g. the number of times the system is in working status), it is possible to estimate the survival function and reliability of the system.

The most generally applicable Monte Carlo simulation methods adopting the survival signature for multi-state component and repairable systems have been proposed in [30]. Its procedural steps are presented in Fig. 7.6.

7.4.1   Systems with Imprecision

The reliability analysis of complex systems requires the probabilistic characterisation of all the possible component transitions. This usually requires a large dataset that is not always available. To avoid the inclusion of subjective assumptions, imprecision and vagueness of the data can be treated by using imprecise probabilities that combine probabilistic and set theoretical components in a unified construct (see, e.g. [4, 9]). Randomness and imprecision are considered simultaneously but viewed separately at any time during the analysis and in the results [32].

Imprecision can occur at component level, where the exact failure distribution is not known or at system level, in the form of an imprecise survival signature. The latter occurs when part of the system can be unknown or not disclosed. Such imprecision leads to bounds on the survival function of the system, providing confidence in the analysis, in the sense that it does not make any additional hypothesis regarding to the available information. When the imprecision is at the component level, a naïve approach, employing a double loop sampling approach where the outer loop is used to sample realisations of component parameters, can be used. In other words, each realisation defines a new probabilistic model that needs to be solved adopting the simulation methods proposed above, from which the envelop of the system reliability is identified. However, since almost all systems are coherent (a system is coherent if each component is relevant, and the structure function is nondecreasing), it is only necessary to compute the system reliability twice, using the lower and upper bounds for all the parameters, respectively. If the imprecision is at the system level (i.e. in the survival signature), the simulation strategy proposed in Fig. 7.6 can be adopted without additional computational cost by collecting, in two separate counters, the upper and lower bounds of the survival signature at each component transition, as illustrated in [30]. Hence, imprecision at the component and system levels can be considered concurrently, without additional computational costs.

7.4.2   Case Study: Industrial Water Supply System

An industrial water supply system consisting of 13 components, as shown in Fig. 7.7, is chosen as a case study, to demonstrate the capability of the survival signature method. The system is expected to deliver water to at least one of the two tanks $T2$ or $T3$ from tank $T1$, through a set of motor-operated pumps and valves. The component failure data with the corresponding distributions are provided in Table 7.2. The survival signature method is employed to compute the reliability of the system.

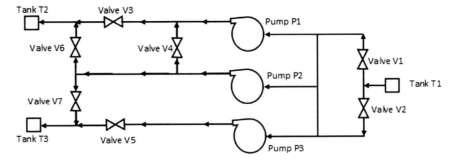

Fig. 7.7 Industrial water supply system

Table 7.2 Reliability parameters of the components of the water supply system

Component	Failure rate (h^{-1})	MTTR (h)	Repair rate (h^{-1})	Distribution type
T_1, T_2, T_3	$\lambda_1 = 5 \cdot 10^{-5}$	24	$\mu_1 = 0.0417$	Exponential
P_1, P_2, P_3	$\lambda_2 = 3 \cdot 10^{-3}$	17.4	$\mu_2 = 0.0575$	Exponential
$V_1, V_2, V_3, V_4, V_5, V_6, V_7$	$\lambda_3 = 2 \cdot 10^{-4}$	9	$\mu_3 = 0.111$	Exponential

Table 7.3 Survival signature (selected parts only) for the system shown in Fig. 7.7 computed with approach proposed in [20]

l_1	l_2	l_3	Φ	l_1	l_2	l_3	Φ
[0, 1]	$\forall$	$\forall$	0				
2	1	2	1/63	3	1	2	1/21
2	1	5	8/63	3	1	5	8/21
2	1	7	2/9	3	1	7	2/3
2	2	6	22/63	3	2	6	6/7
2	3	5	8/21	3	3	5	6/7
2	3	7	2/3	3	3	7	1

The components of the system are categorised into three types, namely, pumps, tanks and valves. The survival signature is given in Table 7.3. The survival function of the water system is then calculated analytically as shown below:

$$P(T_S > t) = \sum_{l_1=0}^{3} \sum_{l_2=0}^{3} \sum_{l_1=0}^{7} \Phi(l_1, l_2, l_3) \binom{3}{l_1} [1 - e^{-\lambda_1 t}]^{3-l_1} \left[e^{-\lambda_1 t}\right]^{l_1} \times$$
$$\binom{3}{l_2} [1 - e^{-\lambda_2 t}]^{3-l_2} \left[e^{-\lambda_2 t}\right]^{l_2} \times \binom{7}{l_3} [1 - e^{-\lambda_3 t}]^{7-l_3} \left[e^{-\lambda_3 t}\right]^{l_3} \quad (7.4)$$

The resulting survival functions without repair and with repairable components are shown in Fig. 7.8.

As shown in Fig. 7.8, the results of the simulation method are in agreement with the analytical solution for both repairable and non-repairable components. The proposed simulation method is applicable to any distribution type, intervals or even probability boxes. It not only separates the system structure from its component failure time distributions, but also doesn't require the *iid* assumption between different component types, as illustrated in [14].

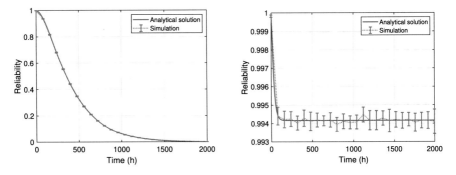

Fig. 7.8 Survival function without repairable (left panel) and with repairable components (right panel) computed using 10000 samples and verified by the analytical solutions

7.5 Final Remarks

System topological complexity, component interdependencies, multi-state component attributes and complex maintenance strategies inhibit the application of simple reliability engineering reasoning to systems. For systems characterised by these attributes, simulation-based approaches allow the realistic analysis of their reliability, despite the relatively higher computational costs of these approaches. This, however, is not a problem, with recent advancement in computing.

The load flow simulation approach is an intuitive simulation framework that is applicable to binary and multi-state systems of any topology. It does not require the prior definition of the structure function, minimal cut sets or the minimal path sets of the system. Instead, it employs a linear programming algorithm and the principles of flow conservation to compute the flow through the system. Thus, it can model flow losses and implement reconfiguration requirements relatively easily. It can model all forms of interdependencies in realistic systems, using intuitive representations. These attributes render the framework intuitive and generally applicable.

While the load flow simulation approach is optimised for multi-state systems, it may not be the best for binary-state systems with identical components. Since the survival signature is a function of the system topology only, it can be calculated only once and reused in multiple reliability analyses. This feature reduces the reliability evaluation of the system to the analysis of the failure probabilities of its components, which is computationally cheap. Efficient simulation methods based on system survival signature allow the reliability analysis of complex systems without resorting to simplifications or approximations.

The load flow and survival signature simulation approaches are not alternative to each other; instead, they can be coupled to take advantage of their unique features, especially for systems with multiple outputs and potentially, multi-state systems.

The algorithms and examples presented are available at: https://github.com/cossan-working-group/SystemReliabilityBookChapter.

References

1. M. Abd-El-Barr. *Design and analysis of reliable and fault-tolerant computer systems*. Hackensack, NJ, USA: World Scientific Publishing Co., 2006.
2. Ahmad W. Al-Dabbagh and Lixuan Lu. Reliability modeling of networked control systems using dynamic flowgraph methodology. *Reliability Engineering and System Safety*, 95(11):1202 – 1209, 2010.
3. L.J.M Aslett, F.P. Coolen, and S.P. Wilson. Bayesian inference for reliability of systems and networks using the survival signature. *Risk Analysis*, 2014.
4. Michael Beer and Edoardo Patelli. Editorial: Engineering analysis with vague and imprecise information. *Structural Safety*, 52:143, 2015.
5. Sergey V. Buldyrev, Roni Parshani, Gerald Paul, H. Eugene Stanley, and Shlomo Havlin. Catastrophic cascade of failures in interdependent networks. *Nature*, 464(7291):1025–1028, April 2010.
6. Sungil Byun, Inseok Yang, Moo Geun Song, and Dongik Lee. Reliability evaluation of steering system using dynamic fault tree. In *IEEE Intelligent Vehicles Symposium*, pages 1416 – 1420, 2013.
7. F.P. Coolen and T. Coolen-Maturi. Generalizing the signature to systems with multiple types of components. *Complex Systems and Dependability*, pages 115 – 30, 2012.
8. Frank P.A. Coolen and Tahani Coolen-Maturi. Predictive inference for system reliability after common-cause component failures. *Reliability Engineering & System Safety*, 135:27 – 33, 2015.
9. Alvarez DA. Infinite random sets and applications in uncertainty analysis. *Arbeitsbereich für Technische Mathematik am Institut für Grundlagen der Bauingenieurwissenschaften. Leopold-Franzens-Universität Innsbruck*, 2007.
10. Salvatore Distefano and Antonio Puliafito. Reliability and availability analysis of dependent-dynamic systems with drbds. *Reliability Engineering and System Safety*, 94(9):1381–1393, 2009.
11. S. Eryilmaz. Review of recent advances in reliability of consecutive k-out-of-n and related systems. *Proceedings of the Institution of Mechanical Engineers, Part O: Journal of Risk and Reliability*, 224 (3):225 - 237, 2010.
12. Samaniego FJ. System signatures and their applications in engineering reliability. *Springer Science & Business Media*, 110.
13. M. Beer F. P. Coolen G. Feng, E. Patelli. Imprecise system reliability and component importance based on survival signature. *Reliability Engineering & System Safety*, 150:116 - 125, 2016.
14. Hindolo George-Williams, Geng Feng, Frank Coolen, Michael Beer, and Edoardo Patelli. Extending The Survival Signature Paradigm To Complex Systems With Non-repairable Dependent Failures. *Proceedings of the Institution of Mechanical Engineers, Part O: Journal of Risk and Reliability*, 233(4):505–519, August 2019.
15. Hindolo George-Williams, Geng Feng, Frank PA Coolen, Michael Beer, and Edoardo Patelli. Extending the survival signature paradigm to complex systems with non-repairable dependent failures. *Proceedings of the Institution of Mechanical Engineers, Part O: Journal of Risk and Reliability*, 233(4):505–519, 2019.

16. Hindolo George-Williams, Min Lee, and Edoardo Patelli. Probabilistic risk assessment of station blackouts in nuclear power plants. *IEEE Transactions on Reliability*, 67(2):494–512, June 2018.

17. Hindolo George-Williams and Edoardo Patelli. A hybrid load flow and event driven simulation approach to multi-state system reliability evaluation. *Reliability Engineering & System Safety*, 152:351–367, August 2016.

18. Hindolo George-Williams and Edoardo Patelli. Efficient Availability Assessment of Reconfigurable Multi-State Systems with Interdependencies. *Reliability Engineering & System Safety*, 165:431–444, September 2017.

19. Hindolo George-Williams and Edoardo Patelli. Maintenance Strategy Optimization for Complex Power Systems Susceptible to Maintenance Delays and Operational Dynamics. 66(4):1309–1330, 2017.

20. M. Beer J. Behrensdorf, M. Broggi. Imprecise reliability analysis of complex interconnected networks. *Safety and Reliability - Safe Societies in a Changing World - Haugen et al. (Eds), Taylor & Francis Group, London, ISBN 978-0-8153-8682-7 year = 2018.*

21. G. Levitin and A. Lisnianski. *Multi-state System Reliability Analysis and Optimization, in: Handbook of Reliability Engineering*, chapter 4, pages 61–90. Springer, 2003.

22. G. Levitin, L. Xing, H. Ben-Haim, and Y. Dai. Multi-state systems with selective propagated failures and imperfect individual and group protections. *Reliability Engineering & System Safety*, 96(12):1657–1666, 12 2011.

23. Jing-An Li, Yue Wu, Kin Keung Lai, and Ke Liu. Reliability estimation and prediction of multi-state components and coherent systems. *Reliability Engineering & System Safety*, 88(1):93 – 98, 2005.

24. A. Lisnianski and Y. Ding. Inverse lz-transform for a discrete-state continuous-time markov process and its application to multi-state system reliability. In *Applied Reliability Engineering and Risk Analysis*. Wiley, 2014.

25. Anatoly Lisnianski, Ilia Frenkel, and Yi Ding. *Multi-State System Reliability Analysis and Optimization for Engineers and Industrial Managers*. Springer-Verlag London Limited, 2010.

26. Manish Malhotra and Kishor S. Trivedi. Dependability modeling using petri-nets. *IEEE Transactions on Reliability*, 44(3):428 – 440, 1995.

27. A Mosleh, D M Rasmuson, and F M Marshall. Guidelines on modeling Commom-Cause Failures in probabilistic risk assessment. Technical Report NUREG/CR-5485, U.S. Nuclear Regulatory Commission, 1998.

28. Andrew O'Connor and Ali Mosleh. A general cause based methodology for analysis of common cause and dependent failures in system risk and reliability assessments. *Reliability Engineering & System Safety*, 145:341 – 350, 2016.

29. Min Ouyang. Review on modeling and simulation of interdependent critical infrastructure systems. *Reliability Engineering & System Safety*, 121:43 – 60, 2014.

30. Edoardo Patelli. *Handbook of Uncertainty Quantification*, chapter COSSAN: A Multidisciplinary Software Suite for Uncertainty Quantification and Risk Management, pages 1–69. Springer International Publishing, 2017.

31. Edoardo Patelli, Hindolo George-Williams, Jonathan Sadeghi, Roberto Rocchetta, Matteo Broggi, and Marco de Angelis. OpenCossan 2.0: An efficient computational toolbox for risk, reliability and resilience analysis. In André T. Beck, Gilberto F. M. de Souza, and Marcelo A. Trindade, editors, *Proceedings of the Joint ICVRAM ISUMA UNCERTAINTIES Conference*, 2018.

32. Broggi M de Angelis M. Patelli E, Alvarez DA. Uncertainty management in multidisciplinary design of critical safety systems. *J Aerosp Inf Syst*, 12:140-69, https://doi.org/10.2514/1. I010273,2015.

33. Jose E. Ramirez-Marquez and David W. Coit. Optimization of system reliability in the presence of common cause failures. *Reliability Engineering & System Safety*, 92(10):1421 – 1434, 2007.

34. S. Reed. An efficient algorithm for exact computation of system and survival signatures using binary decision diagrams. *Reliability Engineering & System Safety*, 165:257 - 267, https://doi.org/10.1016/j.ress.2017.03.036.,2017.

35. Dan M. Shalev and Joseph Tiran. Condition-based fault tree analysis (cbfta): A new method for improved fault tree analysis (fta), reliability and safety calculations. *Reliability Engineering and System Safety*, 92:1231 – 1241, 2007.
36. Masoud Taheriyoun and Saber Moradinejad. Reliability analysis of a wastewater treatment plant using fault tree analysis and monte carlo simulation. *Environmental Monitoring and Assessment*, 187(1), 2015.
37. Matthias CM Troffaes, Gero Walter, and Dana Kelly. A robust Bayesian approach to modeling epistemic uncertainty in common-cause failure models. *Reliability Engineering & System Safety*, 125:13–21, 2014.
38. M. Veeraraghavan and K.S. Trivedi. A combinatorial algorithm for performance and reliability analysis using multistate models. *IEEE Transactions on Computers*, 43(2):229–234, Feb 1994.
39. Liudong Xing and Yuanshun Dai. A new decision-diagram-based method for efficient analysis on multistate systems. *Dependable and Secure Computing, IEEE Transactions on*, 6(3):161–174, July 2009.
40. Wei-Chang Yeh. An improved sum-of-disjoint-products technique for the symbolic network reliability analysis with known minimal paths. *Reliability Engineering & System Safety*, 92(2):260 – 268, 2007.
41. Wei-Chang Yeh. A fast algorithm for quickest path reliability evaluations in multi-state flow networks. *Reliability, IEEE Transactions on*, 64(4):1175–1184, Dec 2015.
42. Wei-Chang Yeh, Yi-Cheng Lin, Y.Y. Chung, and Mingchang Chih. A particle swarm optimization approach based on monte carlo simulation for solving the complex network reliability problem. *Reliability, IEEE Transactions on*, 59(1):212–221, March 2010.
43. Enrico Zio, Piero Baraldi, and Edoardo Patelli. Assessment of the availability of an offshore installation by monte carlo simulation. *International Journal of Pressure Vessels and Piping*, 83(4):312 – 320, 2006.

Chapter 8
Overview of Stochastic Model Updating in Aerospace Application Under Uncertainty Treatment

Sifeng Bi and Michael Beer

Abstract This chapter presents the technique route of model updating in the presence of imprecise probabilities. The emphasis is put on the inevitable uncertainties, in both numerical simulations and experimental measurements, leading the updating methodology to be significantly extended from deterministic sense to stochastic sense. This extension requires that the model parameters are not regarded as unknown-but-fixed values, but random variables with uncertain distributions, i.e. the imprecise probabilities. The final objective of stochastic model updating is no longer a single model prediction with maximal fidelity to a single experiment, but rather the calibrated distribution coefficients allowing the model predictions to fit with the experimental measurements in a probabilistic point of view. The involvement of uncertainty within a Bayesian updating framework is achieved by developing a novel uncertainty quantification metric, i.e. the Bhattacharyya distance, instead of the typical Euclidian distance. The overall approach is demonstrated by solving the model updating sub-problem of the NASA uncertainty quantification challenge. The demonstration provides a clear comparison between performances of the Euclidian distance and the Bhattacharyya distance, and thus promotes a better understanding of the principle of stochastic model updating, as no longer to determine the unknown-but-fixed parameters, but rather to reduce the uncertainty bounds of the model prediction and meanwhile to guarantee the existing experimental data to be still enveloped within the updated uncertainty space.

S. Bi (✉)
School of Aerospace Engineering, Beijing Institute of Technology, Beijing, China
e-mail: sifeng.bi@bit.edu.cn

M. Beer
Institute for Risk and Reliability, Leibniz Universität Hannover, Hanover, Germany
e-mail: beer@irz.uni-hannover.de

© The Author(s) 2022
L. Aslett et al. (eds.), *Uncertainty in Engineering*,
SpringerBriefs in Statistics,
https://doi.org/10.1007/978-3-030-83640-5_8

8.1 Introduction

Computational models of large-scale structural systems with acceptable precision, robustness and efficiency are critical, especially for applications where a large amount of experimental data is hard to be obtained such as the aerospace engineering. Model updating has been developed as a typical technique to reduce the discrepancy between the numerical simulations and the experimental measurements. Recently, it is a tendency to consider the inevitable uncertainties involved in both simulations and experiments. A better understanding of the discrepancy between them, in the background of uncertainty, would achieve a better outcome of model updating.

In structural analysis, the source of uncertainties, i.e. the reason of the discrepancy between simulations and measurements, can be classified into the following categories:

- Parameter uncertainty: Imprecisely known input parameters of the numerical model, such as material properties of novel composites, geometry sizes of complex components and random boundary conditions;
- Modelling uncertainty: Unavoidable simplifications and idealisations, such as linearised representations of nonlinear behaviours and frictionless joint approximations;
- Experimental uncertainty: Hard-to-control random effects, such as environment noise, measurement system errors and human subjective judgments.

The above uncertainties motivate the trend to extend model updating from the deterministic sense to the stochastic sense. The stochastic updating techniques draw massive attention in the literature, in which the majority is based on the framework of imprecise probability [4]. Considering the very typical categorisation of uncertainties, the term "imprecise probability" can be understood separately as "probability" corresponding to the aleatory uncertainty, and "imprecise" corresponding to the epistemic uncertainty. The epistemic uncertainty is caused by the lack of knowledge. As a better understanding of the problem is achieved, this part of uncertainty can be reduced by model updating. The aleatory uncertainty represents the natural randomness of the system, such as the random wind load on launch vehicles, manufacture and measurement system errors. This part of uncertainty is irreducible, however, an appropriate quantification of the aleatory uncertainty is still required in stochastic model updating.

The involvement of aleatory and/or epistemic uncertainties provides a clear logic for the categorisation of model input parameters:

   I Parameters without any aleatory or epistemic uncertainty, appearing as explicitly determined constants;
  II Parameters with only epistemic uncertainty, which are still constants, however, with an undetermined position in an interval;
 III Parameters with only aleatory uncertainty, which are no longer constants, but presented as random variables with exactly known distribution characteristics;

IV Parameters with both aleatory and epistemic uncertainties, which are random variables with undetermined distribution characteristics.

The above parameters with various uncertainty characteristics lead the model predictions into the fourth category, i.e. the outputs with imprecise probabilities. Consequently, the objective of stochastic model updating is no longer a single model prediction with maximal fidelity to a single experiment, but rather a minimised uncertainty space of the outputs, whose bounds should still encompass the existing experimental data. In order to achieve this objective, a novel Uncertainty Quantification (UQ) metric based on the Bhattacharyya distance is introduced in this chapter. The UQ metric refers to an explicit value quantifying the discrepancy between simulations and measurements. Clearly, this metric is expected to be as comprehensive as possible to capture all sources of uncertainty information simultaneously. Furthermore, the overall updating procedure is committed to being simple enough, by employing a uniform framework applicable to both of the classical Euclidian distance and the novel Bhattacharyya distance. Within this uniform framework, comparison between these two distance-based metrics can be performed conveniently, and therefore a better understanding of the difference between the deterministic and stochastic updating is achieved with significantly reduced calculation cost.

The following parts of this chapter are organised as follows. Section 8.2 gives an overview of the state of the art of deterministic and stochastic model updating where key literature is provided. Section 8.3 presents the overall technique route with description of the key aspects, which can be helpful to generate a preliminary figure of the overall model updating campaign. As the emphasis, the parameter calibration procedure with uncertainty treatment is explained in Sect. 8.4, where the Bhattacharyya distance-based UQ metric is presented along with the Bayesian updating framework. The NASA UQ challenge problem is solved by the proposed approach and some interesting results are compared and analysed in Sect. 8.5. Section 8.6 presents the conclusions and prospects.

8.2 Overview of the State of the Art: Deterministic or Stochastic?

Although this chapter focuses on the stochastic model updating, the deterministic updating is still the footstone of its stochastic extension. A comprehensive review of the deterministic model updating techniques can be found from Ref. [14]. The readers are also suggested to refer to the fundamental book by Friswell and Mottershead [15] on this subject covering key aspects including model preparation, vibration data acquisition, sensitivity analysis, error localisation, parameter calibration, etc.

Among the plentiful techniques for deterministic parameter calibration, the sensitivity-based method is one of the most popular approaches based on the linearisation of the generally nonlinear relation between the model inputs and outputs. Mottershead et al. [20] provide tutorial literature for the sensitivity-based updating procedure of finite element models with both demonstrative and industry-scale examples. However, the sensitivity-based method is only valid for typical outputs, e.g. natural frequencies, modal shapes and displacement, of modal analysis or static analysis. Other more complex applications such as strong nonlinear dynamics or transient analysis lead the analytically solved sensitivity to be unpractical. Consequently, the random sampling method, more specifically the Monte Carlo method, attracts more and more interest by providing a direct connection between the model parameters and any output features via multiple deterministic model evaluations. The rapid development of computational hardware makes it possible for large-size samples, from which the statistical information of the inputs/outputs can be precisely estimated [17, 18]. Such Monte Carlo-based methods have been successfully applied in large-scale structures, see e.g. Refs. [10, 16].

The widely used Monte Carlo methods obviously benefit to the research of stochastic model updating, meanwhile, its conjunction with the Bayesian theory further promotes this topic. Beck and Katafygiotis [3] proposed the fundamental framework of Bayesian updating, which was further developed via the Markov Chain Monte Carlo (MCMC) algorithm by Beck and Au [2]. The Bayesian updating framework in conjunction with the MCMC algorithm possesses the advantage to capture the uncertainty information presented by rare experimental data. This approach has been developed as a standard solution of stochastic model updating for different applications such as uncertainty characterisation [22].

Besides the Bayesian interface, other imprecise probability techniques also have considerable potential to be applied in stochastic model updating, such as interval probabilities [13], evidence theory [21], info-gap theory [5] and fuzzy probabilities [19]. For the background of imprecise probability, the comprehensive review by Beer et al. [4] is suggested for an overall understanding of this topic.

8.3 Overall Technique Route of Stochastic Model Updating

The implement of stochastic model updating requires a complete theoretic system including various key steps from the originally developed model, with a series of techniques to define the features, to select and calibrate the parameters, to locate and reduce the modelling errors, and finally to validate the model with independent measurements. Special treatment of uncertainty propagation and quantification promotes the extension of model updating from deterministic domain to stochastic domain. This extension is specifically implemented to key steps such as parameter calibration, model adjustment and validation. Overview of all related steps in model updating is provided in the following subsections.

8.3.1 Feature Extraction

A numerical model cannot be universally feasible for all scenarios with different output features. Here, the feature is defined as the quantity that the engineer wants to predict with the model and it is also dependent on the capacity of the practical experimental set-up. Clearly, different features lead to different sensitivities of parameters, and require different strategies for uncertainty quantification. And therefore the first step of model updating is to define a suitable feature according to the existing experimental setup and the practical application. The most typical features in structural updating are the modal quantities, e.g. the natural frequencies and modal shapes. Multiple orders of frequencies constitute a vector and the absolute mean error between two vectors can be easily utilised to quantify the discrepancy between the simulated and measured data. For modal shapes, the Modal Assurance Criterion (MAC) [1] is the most popular tool to quantify the correlation between two sets of eigenvectors. The continuous quantities are also commonly utilised as features, such as the displacement response in time domain and Frequency Response Functions (FRFs) in frequency domain. Classical techniques to evaluate the difference between two complex and continuous quantities are the Signature Assurance Criterion (SAC) and Cross Signature scale Factor (CSF). Reference [8] provides an integrated application of SAC and CSF for a comprehensive comparison between two FRFs.

8.3.2 Parameter Selection

A sophisticated model of a large structure system always contains a massive number of parameters, which lead to a huge calculation burden and even failure of the updating procedure. Parameter selection is therefore a key step to select or filter parameters according to their significances to the features defined in the first step. The core technique for parameter selection is the sensitivity analysis focusing on a quantitative measurement of the parameter significance. The classical sensitivity analysis technique is the Sobol variance-based method [25]. For a comprehensive knowledge of the global sensitivity analysis inspired by Sobol's method, the well-written book by Saltelli et al. [24] is suggested to the readers. Another extension of Sobol method includes, e.g. the Analysis of Variance (ANOVA) based on the hypothesis testing in probabilistic theory. Reference [7] proposes an integrated application of ANOVA and Design of Experiment (DoE), which provides a significant coefficient matrix containing the complete sensitivity information of a multiple parameter–multiple feature system. When uncertainties are involved, the sensitivity analysis requires extension from the deterministic procedure to the stochastic procedure. The sensitivity of a certain parameter, in the background of uncertainty treatment, can be represented as the degree of how much the uncertainty space of the features is reduced, when the epistemic uncertainty space of the parameter is completely reduced. This requires additional techniques for uncertainty propagation and quantification, which will be addressed in the following parameter calibration step. A comprehensive literature review on the subject of sensitivity analysis can be found in Ref. [26].

8.3.3 Surrogate Modelling

The employment of surrogate models, known as mate-models in some literature, becomes increasingly important along with the sampling-based updating methodologies where a large number of model evaluations are generally required. A surrogate model is a fast-running script between inputs and outputs, which can replace the time-consuming model, e.g. the large finite element model, in the updating procedure. An original input/output sample, i.e. a training sample, generated from the complex model is required to train the surrogate model. Since the surrogate model is proposed to handle the conflict between efficiency and precision, the training sample is expected to be as small as possible, while the precision of the surrogate model according to the original model should be high enough. The typical types of surrogate models include the polynomial function, radial basis function, support vector machine, Kriging function neural network, etc. The selection of a suitable surrogate model type is determined by various aspects such as its efficiency, generality, and nonlinearity. Another technique, in conjunction with the surrogate modelling, is the DoE with the purpose to efficiently and uniformly configure a spatial distribution of the training sample within the whole parameter space. A comprehensive review of the existing techniques of surrogate modelling and DoE can be found in Ref. [27].

8.3.4 Test Analysis Correlation: Uncertainty Quantification Metrics

Test Analysis Correlation (TAC) is the core step of the overall updating procedure, not only because it significantly influences the updating outcome but also because it is the part mostly extended by the uncertainty treatment. TAC refers to the process to quantitatively measure the agreement (or lack thereof) between test measurements and analytical simulations, taking uncertainties into account. It therefore requires a comprehensive metric which is capable of capturing multiple uncertainty sources simultaneously. This chapter proposes UQ metrics under various distance concepts. The Euclidian distance, i.e. the absolute geometry distance between two single points, is probably the most common metric used in deterministic updating approaches. However, it becomes insufficient for stochastic updating where multiple simulations and multiple tests are presented. The Mahalanobis distance is a weighted distance considering the covariance between two datasets. And the alternative Bhattacharyya distance is a statistical distance measuring the overlap between two random distributions. A comprehensive comparison among the three distances in model updating and validation can be found in Ref. [9], where the Bhattacharyya distance is found to be more comprehensive to capture more sources of uncertainty information. In the example section, the Bhattacharyya distance-based UQ metric is applied within a Bayesian updating framework, and the result is compared with the one using the typical Euclidian distance. This work does not address the Mahalanobis distance, since it has been found to be infeasible for parameter calibration, although it contains the covariance information among multiple outputs. Nevertheless, the Mahalanobis distance has the potential to be utilised in model validation as demonstrated in Ref. [9].

8.3.5 Model Adjustment and Validation

After the model parameters are calibrated, the model still needs to pass the validation procedure before it can be utilised in a practical application. The validation procedure contains a series of criteria with increasing requirements: (1) The updated model should predict the existing measurements; (2) The updated model should predict another set of measurement data which is different from the ones used for updating; (3) The updated model should predict any modification of the physical system by making the same modification on the model; (4) The updated model, when utilised as a component of a whole system, should improve the prediction of the whole system model. In the background of uncertainty treatment, the fit between the prediction and the measurement should be assessed by not only the precision but also the stochastic characteristics, e.g. probabilities, intervals and probability boxes. Model adjustment is a procedure to deal with the modelling uncertainty. When the updated model fails to fulfil the validation criteria, or some updated parameters are found to be unphysical, e.g. a minus density value, the modelling uncertainty is too severe to be compensated by calibrating the parameters. The model can be adjusted by increasing the resolution, changing the element type and adding a more detailed geometry description, etc. Another round of parameter calibration is performed for the adjusted model until the validation criterion is found to be fulfilled.

8.4 Uncertainty Treatment in Parameter Calibration

8.4.1 The Bayesian Updating Framework

The typical Bayesian model updating methodology is based on the following Bayes' equation:

$$P(\theta|\mathbf{X}_{exp}) = \frac{P_L(\mathbf{X}_{exp}|\theta)P(\theta)}{P(\mathbf{X}_{exp})} \tag{8.1}$$

with the key elements described as follows:

- $P(\theta)$ is the prior distribution of the parameters, representing the prior knowledge before model updating;
- $P(\theta|\mathbf{X}_{exp})$ is the posterior distribution of the parameters conditional to the existing measurement, i.e. $P(\theta|\mathbf{X}_{exp})$ is the outcome of Bayesian updating;
- $P(\mathbf{X}_{exp})$ is the so-called "normalisation factor" guaranteeing the integration of the posterior distribution $P(\theta|\mathbf{X}_{exp})$ equal to one;
- $P_L(\mathbf{X}_{exp}|\theta)$ is the likelihood function defined as the probability of the existing measurements conditional to an instance of the parameters.

The likelihood represents the probability of the measurement data under each instance of the updating parameters θ. And thus the objective of model updating in the Bayesian background is expressed as: to find the specific instance of the parameters allowing the experimental measurement to possess the largest probability, in other words, allowing the likelihood $P_L(\mathbf{X}_{exp}|\theta)$ reach the maximum. See Chap. 1 for additional background on Bayesian inference.

However, one of the difficulties in Bayesian updating is relative to the normalisation fact $P(\mathbf{X}_{exp})$. The direct integration of the posterior distribution over the whole parameter space is quite difficult in practical application, especially when the number of parameters is large and the distribution format is complex, leading the direct evaluation of the normalisation factor impractical. The well-known MCMC algorithm is popular to solve this difficulty by replacing Eq. (8.1) with

$$P(\theta|\mathbf{X}_{exp}) \approx P_L(\mathbf{X}_{exp}|\theta)^\beta P(\theta) \qquad (8.2)$$

where β is the weighting coefficient fallen within the interval [0, 1]. When β equals to zero, the right part of Eq. (8.2) is the prior distribution; when β equals to one, the right part of Eq. (8.2) converges to the posterior distribution. In the j-th iteration of the MCMC algorithm, random samples are generated from the intermediate distributions with weighting coefficient $\beta_j \in [1, 0]$. In the $(j + 1)$-th iteration, parameter points which lead to higher likelihood are selected from the random samples in the j-th iteration. New Markov chains are generated using the selected parameter points, and thus β_{j+1} is updated. The intermediate distribution converges to the posterior distribution when $\beta_j = 1$. The MCMC algorithm has been developed as a standard tool to stepwise generate samples from very complex target distributions. For the detailed description of the MCMC algorithm, Refs. [2, 11] are suggested to the readers. More applications of this algorithm can be found in the fields from stochastic model updating [6, 22] to structural health monitoring [23]. See Chap. 2 for additional background on Monte Carlo methods.

8.4.2 A Novel Uncertainty Quantification Metric

The MCMC algorithm makes it possible to avoid the evaluation of the normalisation fact in Bayes' equation. However, the evaluation of the likelihood is inevitable and it becomes even more critical along with the uncertainty treatment. In the presence of multiple sets of measurement data, the theoretical definition of the likelihood is

$$P_L(\mathbf{X}_{exp}|\theta) = \prod_{k=1}^{N_{exp}} P(\mathbf{x}_k|\theta) \tag{8.3}$$

where N_{exp} is the number of existing experiments. This equation requires accurate knowledge of the distribution of each measurement data point $P(\mathbf{x}_k|\theta)$. An accurate estimation of the distribution requires a large number of samples, which means a large number of model evaluations.

Clearly, a direct evaluation of Eq. (8.3) leads to a huge calculation cost. This is why the Approximate Bayesian Computation (ABC) becomes increasingly popular in Bayesian applications. Considering the principle of the likelihood function, it is natural to propose an approximate function to replace Eq. (8.3), as long as this approximate function still contains the information of the existing measurement data and an instance of updating parameters. In this work, an approximate likelihood based on the Gaussian function is proposed as

$$P_L(\mathbf{X}_{exp}|\theta) \approx \frac{1}{\sigma\sqrt{2\pi}} \exp\left\{-\frac{d(\mathbf{X}_{exp}, \mathbf{X}_{sim})^2}{2\sigma^2}\right\} \tag{8.4}$$

where $d(\mathbf{X}_{exp}, \mathbf{X}_{sim})$ is the distance between the experimental and simulated feature data. Equation (8.4) serves as an elegant connection between the Bayesian updating framework and the distance concepts. It provides a uniform framework for various distance concepts. The approximate likelihood is applicable for not only the Euclidian distance but also for the Mahalanobis and Bhattacharyya distances, in a uniform updating framework. As explained in Sect. 8.3.4, the Mahalanobis distance is not utilised in this work. The Euclidian distance is evaluated as

$$d_E(\mathbf{X}_{exp}, \mathbf{X}_{sim}) = \left[(\overline{\mathbf{X}_{exp}} - \overline{\mathbf{X}_{sim}})(\overline{\mathbf{X}_{exp}} - \overline{\mathbf{X}_{sim}})^T\right]^{1/2} \tag{8.5}$$

where $\mathbf{X}_{exp}$ and $\mathbf{X}_{sim}$ are the experimental and simulated feature data matrices, respectively; $\overline{\mathbf{X}}$ denotes the mean vector of the matrix. Clearly, the Euclidian distance only handles the mean of the data. The Bhattacharyya distance has the definition as

$$d_B(\mathbf{X}_{exp}, \mathbf{X}_{sim}) = -\log\left[\int_{\mathbb{Y}} \sqrt{P_{exp}(x) P_{sim}(x)} dx\right] \tag{8.6}$$

where $P(x)$ is the PDF of the feature; $\mathbb{X}$ is the feature space, implying $\int_{\mathbb{X}}$ is the integration performed over the whole feature variable space. More detailed information about the evaluation method of the Bhattacharyya distance can be found in Ref. [6].

8.5 Example: The NASA UQ Challenge

The NASA UQ challenge [12] has been developed as a benchmark problem for uncertainty treatment techniques in multidisciplinary engineering. This challenge problem contains a series of subproblems such as uncertainty characterisation, sensitivity analysis, uncertainty propagation, etc. In this work, only the Subproblem A (Uncertainty Characterisation) is investigated since it is equivalent to the task of model updating herein. Figure 8.1 illustrates the key components of this problem, which contains one output evaluated via a black-box model using five input parameters. According to the categorisation strategy in Sect. 8.1, the five parameters are classified into three categories, as shown in Table 8.1.

Only the parameters involving epistemic uncertainties, i.e. Categories II and IV parameters, require to be updated in this context. The uncertainty characteristics, including distribution types and distribution coefficients, are predefined as listed in Table 8.1. The predefined intervals of the distribution coefficients represent the epistemic uncertainty of the parameters. An output sample with 50 data points is available in the problem, which is generated by assigning a set of "true" values of the distribution coefficients from the predefined intervals. The objective of this problem is, based on the existing 50 output points, to reduce the epistemic uncertainty space of the parameters, i.e. to reduce the predefined intervals of the distribution coefficients. Although the number of the model parameter is five, there are totally eight updating coefficients controlling the epistemic uncertainty space of the parameters, as shown in the last a column of Table 8.1.

To solve the problem, the Euclidian and Bhattacharyya distances are utilised to construct the approximate likelihood functions, respectively. The Bayesian updating

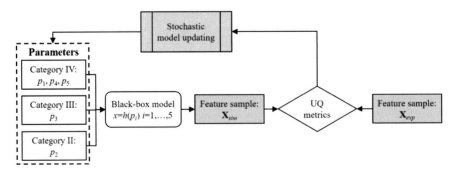

Fig. 8.1 Key components of the Subproblem A in the NASA UQ challenge

Table 8.1 The uncertainty characteristics of the parameters

Parameters	Categories	Uncertainty characteristics	Updating coefficients
p_1	IV	Unimodal Beta, $\mu_1 \in [0.6, 0.8]$, $\sigma_1^2 \in [0.02, 0.04]$	$\theta_1 = \mu_1, \theta_2 = \sigma_1^2$
p_2	II	Constant, $p_2 \in [0.0, 1.0]$	$\theta_3 = p_2$
p_3	III	Uniform, $\mu_3 = 0.5$, $\sigma_3^2 = 1/12$	No updating required
p_4, p_5	IV	Joint Gaussian, $\mu_i \in [-5.0, 5.0]$, $\sigma_i^2 \in [0.0025, 4.0]$, $\rho \in [-1.0, 1.0], i = 4, 5$	$\theta_4 = \mu_4, \theta_5 = \sigma_4^2$ $\theta_6 = \mu_5, \theta_7 = \sigma_5^2$, $\theta_8 = \rho$

framework employs these two distances as UQ metrics and generates two independent sets of results. This treatment is intended to make a clear distinction between the deterministic updating and the stochastic updating, and furthermore to reveal the merits and demerits of these two distances. In practical applications, however, a combined application of these two distances is suggested by first using the Euclidian distance for the mean updating, and second using the Bhattacharyya distance for the variance updating. This two-step strategy has been demonstrated as a success in solving this problem in Ref. [6].

The posterior distributions of the eight updating coefficients using respectively the Euclidian and Bhattacharyya distances as the UQ metrics are illustrated in Fig. 8.2. In the figure, the objects ("Samples" or "PDFs") with the suffix "_ED" denote the

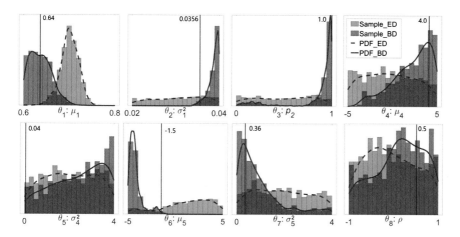

Fig. 8.2 Posterior distributions updated using the Euclidian and Bhattacharyya distances

results updated using the Euclidian distance metric; and the ones with the suffix "_BD" denote the results using the Bhattacharyya distance metric. Except the distribution of θ_1, most of the posterior distributions with the Euclidian distance metric are still close to uniform, implying the deterministic updating procedure employing the Euclidian distance is incapable of solving this problem. As an obvious comparison, the updating procedure employing the Bhattacharyya distance performs well for most of the updating coefficients by providing very peaked posterior distributions, such as θ_2, θ_3, θ_6, and θ_7. The vertical line in each subfigure represents the true value of the updating coefficient, which were used to generate the existing 50 output samples. By comparing the position of the vertical line with the peak of the posterior distribution, the updating precision is assessed. For θ_1, the peak of the distribution with the Euclidian distance is apart from the vertical line, while the peak of the distribution with the Bhattacharyya distance is quite close to the vertical line. This means the Bhattacharyya distance performs better than the Euclidian distance when updating θ_1. The same conclusion is also achieved for θ_3, θ_4, and θ_7. However, for θ_2, and θ_5, although the distributions with the Bhattacharyya distance have clear poles, they do not converge to the vertical lines. The updating precision of these two coefficients is possible to be further improved by the two-step strategy as described in Ref. [6]. Note that, there are still two coefficients (θ_5 and θ_8), which cannot be calibrated by neither the Euclidian distance nor the Bhattacharyya distance. A potential explanation is that the sensitivity of these two coefficients to the output is extremely low, leading the inverse procedure impossible to locate the "true" value.

The quantitative updating results are detailed in Table 8.2, where the true values of the updating coefficients and the updated ones are provided. The updated values in the last two columns of Table 8.2 are obtained by determining the exact position of the distribution peaks in Fig. 8.2. For the posterior distributions which are still close to uniform distribution, the determining process is meaningless and thus omitted. This is why only two updated coefficients are provided in the column with the Euclidian distance. Note that, although the true values of the coefficients are released in Table 8.2, they are not necessarily to be treated as the final target of this updating problem. In the background of uncertainty treatment, the objective of this problem, stipulated by the problem designer [12], is to reduce the epistemic uncertainty space of the parameters, while making sure that the existing output sample can still be included in the output uncertainty space. Hence, it makes more sense to assess how much the output uncertainty space has been reduced after the intervals of the coefficients are reduced in the updating procedure. This is relative to the tasks uncertainty propagation and model validation, which are out of the scope of this example for parameter calibration. A complete validation procedure considering the uncertainty space of the output for this problem can be found in Ref. [6]. Nevertheless, the stochastic Bayesian updating framework employing the Bhattacharyya distance has been demonstrated to be more comprehensive and feasible than the Euclidian distance in solving this NASA UQ challenge problem.

Table 8.2 The updated results using the Euclidian and Bhattacharyya distances

Updating coefficients	Predefined intervals	True values[a]	Updated results	
			Euclidian	Bhattacharyya
θ_1: μ_1	[0.6, 0.8]	0.6364	0.6990	0.6194
θ_2: σ_1^2	[0.02, 0.04]	0.0356	–	0.0397
θ_3: p_2	[0, 1]	1.0	–	0.9855
θ_4: μ_4	[−5, 5]	4.0	–	3.9770
θ_5: σ_4^2	[0.0025, 4]	0.04	–	–
θ_6: μ_5	[−5 , 5]	−1.5	–	−4.4732
θ_7: σ_5^2	[0.0025, 4]	0.36	–	0.2818
θ_8: ρ	[−1, 1]	0.5	−0.3471	0.1804

[a]Data available online at https://uqtools.larc.nasa.gov/nda-uq-challenge-problem-2014 [retrieved in 2017]

8.6 Conclusions and Prospects

The tendency of uncertainty analysis has been rendering the typical model updating full of vitalities but also challenges. This chapter reviews the key techniques and components of the overall model updating campaign. Main emphasis is put on the involvement of uncertainty, which leads the transformation from the deterministic approach to the stochastic approach. The stochastic model updating is executed within the Bayesian model updating framework, where the Bhattacharyya distance is proposed as a novel UQ metric. The approximate likelihood is critical by providing a uniform connection between the Bayesian framework and various types of distance metrics. The example demonstrates that the Bhattacharyya distance is more comprehensive and feasible than the Euclidian distance to calibrate distribution coefficients of parameters with imprecise probabilities.

The Bhattacharyya distance is designed as a universal tool of UQ, which can be conveniently embedded into a technique route similar as the deterministic approach, but provides stochastic outcomes by capturing more uncertainty information. The tendency of uncertainty analysis will be further promoted by the novel UQ metric in not only the stochastic parameter calibration but also other procedures, e.g. the stochastic sensitivity analysis and model validation, which will establish the complete scenario of the stochastic model updating.

References

1. R. Allemang. The modal assurance criterion: Twenty years of use and abuse. *Sound and vibration*, 8:14–21, 2008.
2. J. Beck and S.K. Au. Bayesian updating of structural models and reliability using markov chain monte carlo simulation. *Journal of Engineering Mechanics*, 128(4):380–391, 2002.

3. J. Beck and L. Katafygiotis. Updating models and their uncertainties. i: Bayesian statistical framework. *Journal of Engineering Mechanics*, 124(4):455–461, 1998.
4. M. Beer, S. Ferson, and V. Kreinovich. Imprecise probabilities in engineering analyses. *Mechanical Systems and Signal Processing*, 37(1-2):4–29, 2013.
5. Y. Ben-Haim. *Info-gap decision theory: Decisions under severe uncertainty*. Elsevier, 2006.
6. S. Bi, M. Broggi, and M. Beer. The role of the bhattacharyya distance in stochastic model updating. *Mechanical Systems and Signal Processing*, 117:437–452, 2019.
7. S. Bi, Z. Deng, and Z. Chen. Stochastic validation of structural fe-models based on hierarchical cluster analysis and advanced monte carlo simulation. *Finite Elements in Analysis and Design*, 67:22–33, 2013.
8. S. Bi, M. Ouisse, and E. Foltête. Probabilistic approach for damping identification considering uncertainty in experimental modal analysis. *AIAA Journal*, 56(12):4953–4964, 2018.
9. S. Bi, S. Prabhu, S. Cogan, and S. Atamturktur. Uncertainty quantification metrics with varying statistical information in model calibration and validation. *AIAA Journal*, 55(10):3570–3583, 2017.
10. A. Calvi. Uncertainty-based loads analysis for spacecraft: Finite element model validation and dynamic responses. *Computers & Structures*, 83(14):1103–1112, 2005.
11. J. Ching and Y. Chen. Transitional markov chain monte carlo method for bayesian model updating, model class selection, and model averaging. *Journal of Engineering Mechanics*, 133(7):816–832, 2007.
12. L. Crespo, S. Kenny, and D. Giesy. The nasa langley multidisciplinary uncertainty quantification challenge. In *16th AIAA Non-Deterministic Approaches Conference*.
13. M. Faes, M. Broggi, E. Patelli, Y. Govers, J. Mottershead, M. Beer, and D. Moens. A multivariate interval approach for inverse uncertainty quantification with limited experimental data. *Mechanical Systems and Signal Processing*, 118:534–548, 2019.
14. M. Friswell and J. Mottershead. Model updating in structural dynamics: a survey. *Journal of Sound and Vibration*, 162(2):347–375, 1993.
15. M. Friswell and J. Mottershead. *Finite Element Model Updating in Structural Dynamics*. Kluwer Academic Press, Dordrecht, Netherlands, 1995.
16. B. Goller, M. Broggi, A. Calvi, and G.I. Schueller. A stochastic model updating technique for complex aerospace structures. *Finite Elements in Analysis and Design*, 47(7):739–752, 2011.
17. Y. Govers and M. Link. Stochastic model updating - covariance matrix adjustment from uncertain experimental modal data. *Mechanical Systems and Signal Processing*, 24(3):696–706, 2010.
18. C. Mares, J. Mottershead, and M. Friswell. Stochastic model updating: Part 1 - theory and simulated example. *Mechanical Systems and Signal Processing*, 20(7):1674–1695, 2006.
19. B. Möller and M. Beer. *Fuzzy Randomness: Uncertainty in Civil Engineering and Computational Mechanics*. Springer, 2004.
20. J. Mottershead, M. Link, and M. Friswell. The sensitivity method in finite element model updating: A tutorial. *Mechanical Systems and Signal Processing*, 25:2275–2296, 2011.
21. W. Oberkampf and J. Helton. Evidence theory for engineering applications. In *Engineering design reliability handbook*, chapter 10, pages 197–226. CRC Press, 2004.
22. E. Patelli, D. Alvarez, M. Broggi, and M. De Angelis. Uncertainty management in multidisciplinary design of critical safety systems. *Journal of Aerospace Information Systems*, 12(1):140–169, 2015.
23. R. Rocchetta, M. Broggi, Q. Huchet, and E. Patelli. On-line bayesian model updating for structural health monitoring. *Mechanical Systems and Signal Processing*, 103:174–195, 2018.
24. A. Saltelli, M. Ratto, T. Andres, F. Campolongo, J. Cariboni, D. Gatelli, M. Saisana, and S. Tarantola. *Global Sensitivity Analysis. The Primer*. John Wiley & Sons, 2008.
25. I. Sobol. Sensitivity estimates for nonlinear mathematical models. *Mathematical Modeling and Computational experiment*, 1(4):407–414, 1993.
26. P. Wei, Z. Lu, and J. Song. Variable importance analysis: A comprehensive review. *Reliability Engineering and System Safety*, 142:399–432, 2015.

27. R. Yondo, E. Andrés, and E. Valero. A review on design of experiments and surrogate models in aircraft real-time and many-query aerodynamic analyses. *Progress in Aerospace Sciences*, 96:23–61, 2018.

Chapter 9
Aerospace Flight Modeling and Experimental Testing

Olivier Chazot

Abstract Validation processes for aerospace flight modeling require to articulate uncertainty quantification methods with the experimental approach. On this note, the specific strategies for the reproduction of re-entry flow conditions in ground-based facilities are reviewed. It shows how it combines high-speed flow physics with the hypersonic wind tunnel capabilities.

9.1 Introduction

Space missions are built upon massive technology knowledge and on the latest progress in engineering. They fascinate as they represent for the public the most advanced knowledge as well as a typical dive into the unknown. For scientists and engineers, such missions are an occasion to push the scientific knowledge and to establish better what we know to offer solid basis for further discoveries.

In practice, space exploration leads to extreme challenges as it aims to investigate planets, or asteroids, in the solar system and return samples for analysis across very severe flight conditions. Such missions need to be designed using physical models and robust numerical methods. However, those tools used by aerospace engineers remain, for most of them, on the need for more research development for their validation and consolidation to be able to plan successful and fruitful missions.

Aerothermodynamic testing is one of the crucial points for the design of aerospace vehicles. At first place, it aims at establishing as much as knowledge possible on critical flight phenomena. Ground-based facilities are operated to reproduce flight-relevant environment for the testing of the vehicle configuration and its Thermal Protection System (TPS) to allow for an accurate evaluation of their performances. Two types of facilities are classically used for the ground testing to support the pre-flight analysis. First, the required flow-field is reproduced for its analysis in

O. Chazot (✉)
Aeronautics and Aerospace Department, von Karman Institute for Fluid Dynamics, Sint-Genesius-Rode, Belgium
e-mail: chazot@vki.ac.be

L. Aslett et al. (eds.), *Uncertainty in Engineering*,
SpringerBriefs in Statistics,
https://doi.org/10.1007/978-3-030-83640-5_9

high-enthalpy facilities such as shock tunnels or expansion tubes. Then in a second step, the interaction between the dissociated gas and the vehicle' surface is studied to determine the thickness of Thermal Protection Material (TPM) from databases built in plasma wind tunnels. Accurate flight duplication is thus necessary in order to properly address those stringent requirements without over-sizing the TPS.

All the data produced are processed to define at best the modeling for all the required physical phenomena. Very performant frameworks for such processing are the Uncertainty Quantification (UQ) methods. Bayesian approach, in particular, are extremely powerful as they allow to determine the required information from a limited experimental knowledge with a probabilistic treatment. However, such key results, for hypersonic flight, are not only due to powerful mathematical treatment but also thanks to consistent experimental methodologies. This combination between mathematical and experimental approach is essential to generate useful knowledge on the physical modeling to be determined. Then it is primordial to have a good understanding on how the ground testing are linked to the fundamental mathematical models used for simulating the high-speed flow physics. It serves to setup the best experimental environment to allow a fruitful processing of the acquired data.

Therefore, this note intends to review the rationale of the ground testing methodology for aerospace vehicle. It offers a synthesis on the high-speed ground testing underlying the links to their scientific basis. It presents how similitude laws could be applied or need to be adapted as more and more physical phenomena have to be considered for aerospace flights. It would help engineers and applied mathematicians for working together facing the challenges of high-speed flight investigations for aerospace development.

9.2   Aerospace Flights and Planetary Re-entry

Aerospace flight can be considered for very different trajectories. It could follow a planet orbit up to the limit of its escape velocity, but it could also correspond to an interplanetary transfer with super-orbital velocities. Those situations involve a large variety of kinetic energy and trajectories. Figure 9.1 illustrates those typical orbit trajectories, around Earth, with their corresponding velocities.

Super-orbital atmospheric re-entry, also known as hyperbolic re-entry, is characterized by high velocities and is encountered when a probe is entering an atmosphere from a hyperbolic orbit rather than from an elliptical orbit. It is the case for some interplanetary probes or sample return missions. Typical velocities for probes entering the Earth's atmosphere from hyperbolic orbits scale from 11 to 14 km/s, which correspond to specific enthalpies between 60 and 100 MJ/kg. This is considerably higher than the usual re-entry velocities from circular or elliptic orbits, e.g., 8.2 km/s for the Space Shuttle. Up to now, the Stardust probe was the fastest artificial object to perform a controlled re-entry in the Earth's atmosphere, at 12.8 km/s.

The environment of a high-speed re-entry flight is much more severe compare to a Low Earth Orbit (LEO) entry with particularly high heat flux and important stagnation

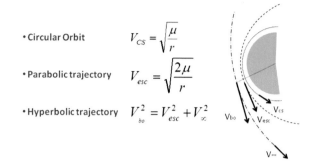

Fig. 9.1 Orbit trajectories and velocities. μ is the Earth gravitational parameter and r is the radius of the orbit

- Circular Orbit $V_{CS} = \sqrt{\dfrac{\mu}{r}}$

- Parabolic trajectory $V_{esc} = \sqrt{\dfrac{2\mu}{r}}$

- Hyperbolic trajectory $V_{bo}^2 = V_{esc}^2 + V_{\infty}^2$

pressure. In addition to the classical features of re-entry flows like non-equilibrium thermo-chemistry and complex gas–surface interactions some specific phenomena become important as shock layer radiative heating and ablation phenomena. Those physical process were not considered much at smaller velocity but they cannot be anymore neglected for super-orbital re-entry. In top of this, more complex physical reality coupling phenomena appear in the flow between radiation and ablation that lead to a very intricate situation. It is therefore not possible to extrapolate what has been studied and learned for orbital re-entry to super-orbital conditions and specific ground testing strategies are required.

9.3   Similitude Approach for Hypersonic Flows

Similitude in classical fluid dynamics establishes a correspondence between different flows, based on the mathematical model representing these flows, without having to solve the set of equations. This correspondence then can be used to relate two real physical flow situations or to relate a family of solutions for the model.

With two correlated applications:

- The flow fields are similar (i.e., solution of the same set of equations) even if the dimensions and the temporal evolution of the phenomena are on different scales.
- When applicable, the use of the similarity laws allows to replicate in wind tunnels the flow field occurring in flight around a re-entry vehicle.

This second aspect of the similitude approach is mostly useful since it allows to study on ground typical re-entry situations. Hypersonic flows are particularly interesting for the application of similitude approach because it exists many different situations in hypersonic regime which can be describes with a variety of models leading to different similitude laws. On one hand, it gives opportunities to develop simplify solutions, but on the other hand, the physical nature of the high-speed flows, mostly due to the high-temperature effects, severely restrict exact similitude and impose to study approximate similitude. This family of flows is essentially determined by the mathematical model chosen to describe the flows and of which the flows are themselves solutions. It appears that the similitude strategy evolves as the model integrates more and more physical aspects for the description of the flow. In hypersonics going from inviscid regime to high-temperature effects, the similitude approach will be adapted to retain the most relevant flow parameters corresponding to each case. Similitude in hypersonic has been studied and discussed by different authors: in earlier time, by Hayes and Probstein for inviscid and viscous hypersonics [11, 12], by Freeman for dissociated gases [9] and in a more general form by Viviand for CFD development [6], but it provides also very useful material for experimental studies. The most common approaches are briefly presented below, as well as their limitations.

9.3.1 Inviscid Hypersonics

At high Mach number when considering a slender body (small thickness ratio $\tau \ll 1$) at small angle of attack ($\alpha \ll 1$) in a perfect gas, the governing Euler equations can be further simplified using the small disturbances theory. The similarity parameters are then $M\tau, \alpha\tau$, and γ. The slenderness ratio τ is defined as $\tau = d/l$, where d is the body's radius and l its length. The parameter $K = M\tau$ is called the hypersonic similarity parameter [11]. On these conditions, with a constant isentropic exponent γ, the results of the similitude may be expressed in dimensionless form. For family of affinely related bodies of thickness ratio, τ in two-dimensional flows the pressure coefficient C_p could be written:

$$\frac{C_p}{\tau^2} = 2\left[\frac{\gamma+1}{4} + \sqrt{\left(\frac{\gamma+1}{4}\right)^2 + \frac{1}{K^2}}\right] = f(K, \gamma) \tag{9.1}$$

The typical testing strategy in this inviscid framework could be represented as in the figure (Fig. 9.2).

That approach holds for inviscid hypersonic, over a wide range of K for slender bodies as long as the Mach number in the flow is large enough and τ small. Nevertheless, it does not correlate well as the body thickness is increased, curved shock and boundary layers start to be predominant in the flow. As most hypersonic vehicles are blunt rather than slender for thermal considerations this hypersonic similarity has limited applications.

9.3.2 Viscous Hypersonics

Hypersonic flows present thick boundary layers that are also growing fast as Mach number is increased. These boundary layers cannot be ignored generally in hypersonic problems as they determine the major feature of the flow physics. They also

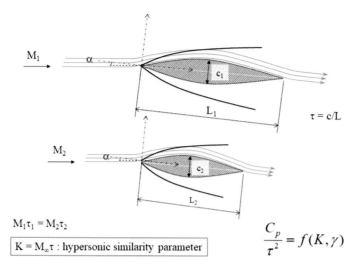

$$M_1\tau_1 = M_2\tau_2$$

$$\boxed{K = M_x\tau : \text{hypersonic similarity parameter}}$$

$$\frac{C_p}{\tau^2} = f(K, \gamma)$$

Fig. 9.2 Inviscid hypersonic similitude

need to be taking into account for their interaction with the inviscid flow; ground-based facilities need to be designed toward those considerations to provide a way for studying these effects in relevant hypersonic flight conditions. To this end, similitude approach is a precious guide to identify which combination of flow parameters need to be taken into account in experimental simulation. Viscous hypersonic similitude has been presented first by Hayes and Probstein in a paper [12] where they review the general features of viscous similitude at high speed. The inviscid flow need to be represented, then the hypersonic similarity parameter $K = M\tau$ and γ need to be invariant. The interaction of the viscous part of the flow with the inviscid field will be determined by the viscous-inviscid interaction parameter χ expressed as

$$\chi = \frac{M_\infty^3 \sqrt{C_\infty}}{\sqrt{Re_{x,\infty}}} \tag{9.2}$$

with $C_\infty = \frac{\mu_r T_\infty}{\mu_\infty T_r}$ and $Re_{x,\infty} = \frac{\rho_\infty U x}{\mu_\infty}$.

With these conditions the perfect gas model has to be assumed and the continuum hypothesis valid, i.e., the mean free path is at least one order of magnitude smaller than the characteristic length of the flow. The viscous hypersonic similarity requires reproducing the free-stream Reynolds (Re) and Mach number (M) and the temperature ratio $\frac{T_w}{T_\infty}$, where the subscript w indicates the wall temperature and the ∞ symbol in the subscript refers to the free-stream. If the gas mixture is not the same, the heat capacity ratio γ also needs to be reproduced [13]. This eases facility development, as lowering the gas temperature lowers the speed of sound, and thereby increases the achievable free-stream Mach number. The condensation temperature of the test

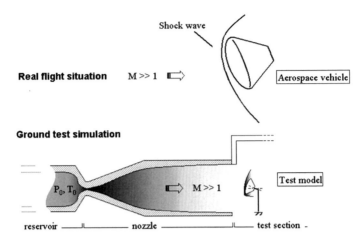

Fig. 9.3 Schematic representation of cold hypersonic testing in ground facility

gas imposes the upper limit of what is achievable in a given facility. In a general view it resorts a Mach–Reynold simulation. Most of the ground testing capabilities are designed following this principle working with scaled models (Fig. 9.3). Their operation envelope are commonly presented in a Ma–Re graph that indicates which flight domain can be simulated (Fig. 9.4).[1] One has to remark that the temperature ratio parameter $\frac{T_w}{T_\infty}$ is not often taking into account but it could be an important aspect to consider in ground testing as it could appear as a limitation in the study.

9.3.3 High-Temperature Hypersonics

When a real gas, as air, experiences a strong shock at high speed, it will increase tremendously its thermal energy. In such conditions, the molecular collisions are energetic enough to cause dissociation and ionization and the gas to depart from the perfect gas model. For a blunt body at a velocity of 7 km/s, the temperature immediately after the shock is around 14,000 K, and around 8,000 K downstream the shock, where the flow may return to equilibrium. At such high temperatures, chemical effects have to be taken into account. For air at a pressure of 1 atm, vibrational excitation begins at 800 K, O2 begins to dissociate at 2,500 K and is fully dissociated for 4,000 K, point for which N_2 begins to dissociate. At 9,000 K, N_2 is fully dissociated and ionization begins. One can easily understand that the flow downstream the shock becomes a plasma: molecules are dissociated and atoms are partially ionized. The parameters to be respected for a flow involving chemistry are $V^2/2D$, where V is the free-stream velocity and D the typical dissociation energy of the gas molecule considered, the Damköhler number Da, defined as $Da = L/l_D$,

---

[1] Graphic extracted from AGARD AR 319.

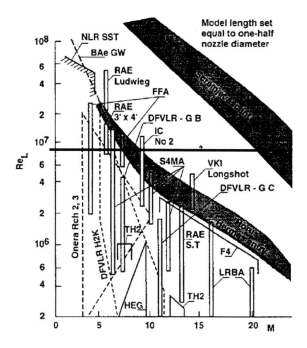

Fig. 9.4 Mach–Reynolds map for hypersonic facilities [20]

where L is the characteristic length of the flow and l_D the characteristic length associated to the dissociation reaction, and the temperature ratio T_w/T [13]. It should be brought to the reader's attention that the gas behind the shock is a chemically reacting mixture of perfect gases, and not a real gas as it is sometime incorrectly referred to in literature.

The Damköhler number for the gas appears considering the mass conservation equation in a non-dimensional form as it is expressed in the equation below:

$$\frac{\partial \rho_i}{\partial t} + \nabla.(\rho_i V + J_i) = \sum_{r=1}^{N_r} (v_{ir}'' - v_{ir}')\left\{ \mathrm{Da}_{fr} \prod_{j=1}^{N_s} \rho_j^{v_{ir}'} - \mathrm{Da}_{bw} \prod_{j=1}^{N_s} \rho_j^{v_{ir}''} \right\} \quad (9.3)$$

$$\mathrm{Da}_{fr} = \frac{k_{fr}\rho_\infty L}{U_\infty} \qquad \mathrm{Da}_{bw} = \frac{k_{bw}\rho_\infty^2 L}{U_\infty} \quad (9.4)$$

Chemical reactions take a very short but finite time to happen. Assuming that the flow is composed of a single species, the dissociation rate is proportional to the density, while the recombination rate is proportional to the square of density. Hence, the characteristic lengths associated to the dissociation and recombination reactions, respectively, scale as $l_D \propto 1/\rho$ and $l_R \propto 1/\rho^2$, where ρ is the flow density [13]. The recombination length is usually larger than the dissociation length. Under certain conditions, at very high altitude, one can assume that $l_D \sim L$, and

Fig. 9.5 High-enthalpy facility map [15]

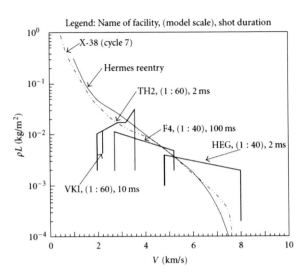

therefore $Da = o(1)$, while l_R is much larger. The flow is frozen; it is too fast for recombination reactions to take place. In that case, the binary scaling parameter ρL must be reproduced in order to obtain the correct Da [16]. One could note that the diffusion phenomena is also considered in this analysis, as the diffusion term in the equation scale with ρL [5]. With this approach, the high-enthalpy facilities are designed to reproduce the real flight velocities (V_∞) and to allow for an operation considering the ρL parameter as they can be represented in the graph of Fig. 9.5.

This leads to some complications. Firstly, the same air mixture as in real flight is commonly used, as the chemistry processes are too complicated to reproduce to use another one. The typical dissociation energy D is therefore conserved, and the actual free-stream velocity must be reproduced to duplicate the group $V^2/2D$. Secondly, the required density to achieve in the wind tunnel becomes large in order to maintain the proper value of the binary scaling parameter for duplication of flight at lower altitude. Thirdly, the binary scaling approach, strictly speaking, is built upon the hypothesis of a single species mixture. That approach has therefore limited application for more complex mixtures such as air [7]. Finally, as altitude decreases, density increases and both l_D and l_R become smaller. The binary scaling parameter does not hold anymore, as both ρL and $\rho^2 L$ should be reproduced at the same time. The same holds when flow velocity increases, and therefore also temperature. This prevents from using the similarity laws, and flow duplication can thus only be performed on a full-scale model, with the actual flow velocity.

9.4   Duplication of Dissociated Boundary Layer with Surface Reaction

Following the development presented in the previous paragraph, it is observed that in spite of all the possibilities offer by the ground testing, considering the high-temperatures effects in hypersonic, only the dissociation in the shock layer could be simulated to some extend on scaled models. If one wanted to further take into account, the recombination in the gas he would have to consider full-scale models. Looking into more details, one could underline that all the important phenomena, concerning the heat transfer typically, are laying in the boundary layer. Figure 9.6 gives an illustration of real flight situation in front of an Aerospace vehicle.

The study could then be reduced to this confined layer in particular for the heat-transfer problems in hypersonic. In this situation, the full-scale environment will be reproduced by duplicating all the characteristics of the boundary layer in the ground testing facilities. This statement was already made in earlier publications from researchers working on dissociated boundary layers using shock tube facilities [19].

The stagnation point is of particular interest for this duplication because the flow conditions, returning to zero velocity, are more easily reproduced in a laboratory. Before identifying the parameters that need to be retained for the testing conditions it will be useful to give a physical description of reacting boundary layers, as they manifest themselves along surfaces in hypersonic flows. Dissociated non-equilibrium boundary layers with surface reaction have already been presented extensively by major authors [2, 19] and the reader is encouraged to consult these references. The

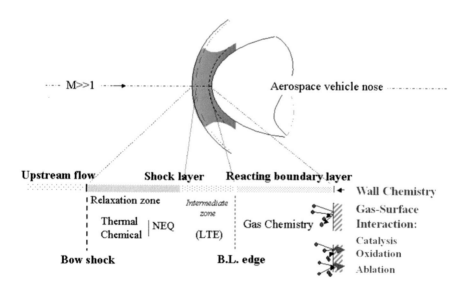

Fig. 9.6 Re-entry environment along stagnation line

purpose of this section is not to expose in details the theory specific to such boundary layers but to recall their main characteristics and better understand how they could be simulated in ground-based facilities. Boundary layers are the location where the diffusion phenomena are dominant, their relative importance are characterized by typical non-dimensional numbers: Prandtl (Pr), Lewis (Le), and Schmidt (Sc). Those will not be commented here to rather focus on the chemical non-equilibrium feature of the flow. The chemical reactions taking place in the gas phase are called homogeneous reactions while those happening between the gas and the solid surface are called heterogeneous reactions. The chemical non-equilibrium in the gas phase is characterized by the gas Damköhler number (Da_g). It corresponds to the ratio between the typical time of the flow, to cross the boundary layer, to a typical reaction time for the gas chemistry: $Da_g = \tau_f/\tau_c$. When $Da_g \to \infty$ the boundary layer reaches a local thermodynamic equilibrium (LTE). On the contrary when $Da_g \to 0$ it leads to a frozen boundary layer, where no chemistry happen. These different conditions have significant consequences on the wall heat flux as it has been shown in reference [8]. The reactions at the wall are usually considered as first-order reactions, they are represented by a reaction rate (k_w) for each dissociated specie. k_w could be related to a recombination coefficient γ_i (or catalycity parameter), interpreted as a probability of recombination at the wall, by the Hertz–Knudsen formula:

$$k_w i = \gamma_i \sqrt{\frac{k \cdot T_w}{2\pi \cdot M_i}} \qquad (9.5)$$

The wall reaction rates are also characterized by a surface Damkohler number (Da_w). It compares the time of diffusion for the species across the boundary layer to the time of reaction at the wall: $Da_w = \tau_{Diff}/\tau_{React}$. When $Da_w \to \infty$, it does not necessarily imply that the reaction rate at the wall tend to infinity ($kw \to \infty$), but simply that the surface reaction are much more faster than the diffusion process. It is said that the GSI phenomena are "limited by diffusion". In the other extreme case, when $Da_w \to 0$ the diffusion is much faster than the reaction and the GSI phenomena are "limited by reaction" or "reaction controlled". From this description, it appears that the diffusion and reaction processes should be accurately reproduced. To this extend, it could be understood that the dimension and the environment of the reaction boundary layer must be duplicated (Fig. 9.7). Two situations could be distinguished for the boundary layer to be duplicated in the laboratory: stagnation point region and off-stagnation point when the boundary is developing along the surface.

If one considers only the stagnation point region, it has been shown that a complete duplication of real flight condition is possible in ground facility, if the total enthalpy (He), the total pressure (Pe), and the velocity gradient ($\beta = du/dx$), of the flight conditions, can be matched locally on the test sample [4, 14]. In this case, the testing is realized in plasma wind tunnels (Arcjet or Plasmatron facilities) that are able to produce dissociated flows for a long time base which is suitable for tests involving aerothermochemistry. The theoretical frame for the testing with ICP wind tunnels has been adapted by Russian scientists, in a methodology called Local Heat Transfer

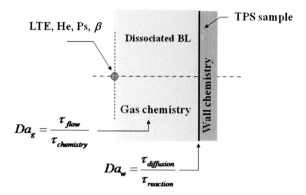

Fig. 9.7 Typical features of the dissociated stagnation point boundary layer

Simulation (LHTS) [14] and an assessment of this methodology has been conducted at the VKI Plasmatron facility [1, 4]. The duplication of the flight condition at stagnation point is strictly reduced to the boundary layer with its appropriate treatment [1]. In the case of a subsonic plasma facility, like the VKI Plasmatron, the testing configuration could be presented as in Fig. 9.8.

The experimental assessment of the LHTS methodology has been conducted at VKI by Chazot et al. [4] and Barbante and Chazot [1]. The results are presented on the Figs. 9.9 and 9.10. It could be seen that the boundary layers profiles from the hypersonic flow is very well duplicated with the subsonic plasma flow, when matching the characteristics parameters at the edge of the boundary layer.

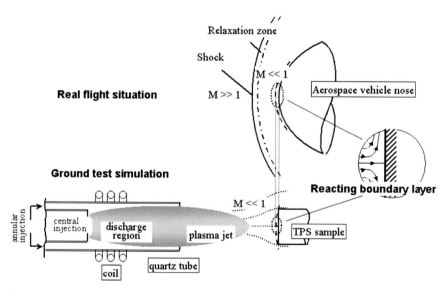

Fig. 9.8 TPS testing in plasma wind tunnel in LHTS conditions

Fig. 9.9 Temperature profiles comparison between flight and ground testing conditions [1]

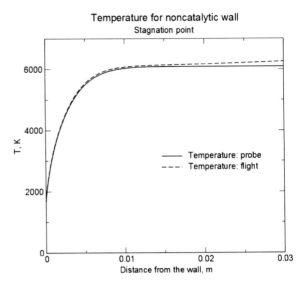

Fig. 9.10 Mass fraction profiles comparison between flight and ground testing conditions [1]

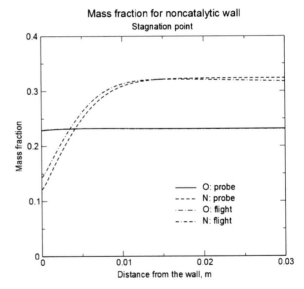

9.5   Considering Flow Radiation

Shock layer radiative heating appears around 9 km/s in Earth's atmosphere and 7 km/s in Mars' atmosphere [17]. It reaches 10 % of the total heat flux for probes having a diameter smaller than 1 m and entry velocities approaching 13 km/s in the Earth's atmosphere [21]. It is then a physical phenomena to be considered for super-orbital re-entry.

However on ground testing for scaled models flow radiation is not often addressed in the literature. It is usually mentioned to explain that this feature of the flow cannot be properly reproduced considering similarity rules for a model of reduced dimension [18]. But even if this conclusion is fully valid some more details could be added to better understand the issues with radiative heat-transfer in hypersonic facilities.

The radiative flux at a certain point P is the integral of the radiative intensity in all directions and over the entire frequency spectrum. Considering a point S on a surface, the radiative flux reaching S from a point P of the surrounding is the difference between the energy emitted and that absorbed integrated along the optical path from P to S. In the case of a scaled model in a high-enthalpy facility, respecting the same condition for the gas and the flow velocity, it could be shown that the optical thickness in the radiating shock layer scales with the product ρL, considering the absorption coefficient remaining the same on the two situations.

One is then brought back to the same approach as the binary scaling exposed before. In these conditions, the radiative heat-flux on the surface of a scaled model could be reproduced in a ground-based facility.

However, if one considers the flow passing through the radiating environment around the model, it appears that the scaling does not hold any longer. The amount of energy E radiated by a control volume is proportional to the mass contained in that volume: $E \propto m = \rho.L^3$. The amount of flow $\dot{m}$ ingested in the shock layer could be expressed as $\dot{m} \propto \rho.U_\infty.L^2$. Therefore, when the flow-radiation coupling need to be taken into account the heat radiated per unit mass passing in a control volume scales as: $E/\dot{m} \propto L$. In conclusion, even if the radiative flux on the surface of a scaled model could be reproduced, the radiative heating of the flow around the same model is not respected.

This problem arises if the radiative heating is important enough to have an influence on the rest of the flowfield. This coupling is quantified by the radiative cooling parameter Γ, also referred to as Goulard number, defined as

$$\Gamma = \frac{2 \cdot Q_{ad}^r}{1/2 \cdot \rho_\infty \cdot U_\infty^3} \tag{9.6}$$

where Q_{ad}^r is the radiative heating for an adiabatic flow, that is without radiative cooling, ρ_∞ the free-stream density, and v the shock velocity. This parameter serves as an approximate measure of the coupling between radiation and the flow [10]. It is the ratio of the amount of radiation generated by the shock, assumed to be twice the radiative heating of the surface, by the kinetic energy heat flux entering the shock layer. If $\Gamma > 0.01$, radiation is coupled to the rest of the flowfields and the effect of improper radiation scaling extends to the rest of the flowfield.

Furthermore, there is a coupling between radiation and gas–surface interaction processes. Indeed, the use of ablative material is compulsory for high-speed re-entry. The ablation processes are very efficient to prevent the hot shock layer gas from reaching the wall and absorb part of the shock layer radiative heat flux. An accurate estimation of the absorbed radiation is complex since the thickness and thermochemical state of the ablation gas layer are difficult to predict up to now [17].

Other surface phenomena such as catalycity and oxidation should be also taken into account, it makes the problem even more intricate when one is aware that such a models are not yet fully established, especially in high-enthalpy flows [3].

9.6 Ground Testing Strategy for High-Speed Re-entry

Ground testing facilities commonly used to study re-entry flows and that are mainly concerned with high-enthalpy flows can be divided in two categories:

- Impulse facilities, such as shock tubes and ballistic ranges, are only able to produce flows that last typically a fraction of a second. It is usually assumed long enough to let the steady flow establish itself, but too short compared to the thermal inertia of the material surface. They are thus mainly used to investigate the aerothermo-dynamic effects, gas kinetic, and radiation processes. In this category, expansion tubes are able to reach free-stream enthalpies characteristic of high-speed re-entry.
- Plasma wind tunnels, such as inductively coupled plasma facilities or arc-jets, are able to operate for long test durations, in the order of minutes. However, they have not been designed to reproduce the flow radiation.

Similar flights conditions or direct flight duplication are possible to reproduce for sub-orbital re-entry velocity in those facilities for a limited time (impulse facilities) or in a limited region (stagnation point in plasma wind tunnels). Each category of facility is addressing a specific aspect of the flowfield as it is sketched in Fig. 9.11.

The velocity of the flow is much higher in the case of high-speed re-entry which concern super-orbital conditions. This results in considerably higher free-stream

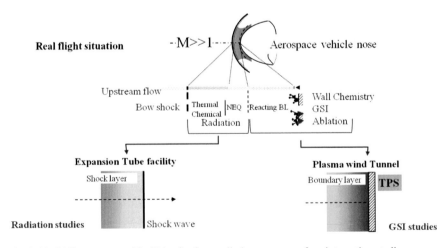

Fig. 9.11 Different types of facilities for flow radiation or gas–surface interaction studies

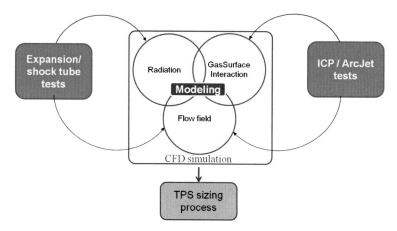

Fig. 9.12 Testing methodology for high-speed re-entry TPS sizing

enthalpy and pressure in the flow, and leads to coupling phenomena that cannot be reproduced on scaled down models.

High-speed re-entry requires therefore a new approach of ground testing. Ground testing facilities should be used to investigate separately different phenomenon playing a role in the aerothermodynamic of the flow, rather than to duplicate flight, as it is the case for lower velocity re-entry.

Those investigations should be performed on a panel of different facilities in order to develop models and databases. Models of shock layer radiation can be developed based on measurements performed in impulse facilities, under conditions similar to those encountered in high-speed re-entry. However, material processes such as ablation and radiative heating cannot be performed in the same facility due to the short test duration involved. In particular, gas–surface interaction have to be studied in plasma wind tunnels. As it is known, those facilities could produce the required heat flux level, but have not been designed to take into account the correct coupling phenomena including the radiation processes present below a shock layer. Models of gas–surface interaction have therefore to be developed under different conditions than that of high-speed re-entry.

Those models, developed separately, should then be implemented in Computational Fluid Dynamics (CFD) codes and allow extrapolatation to the actual flight conditions. Since they cannot be validated within the flight conditions envelope unless flight-testing is performed, they need to capture the main physical phenomena and their accuracy is of prime importance. This, in turn, allows sizing and designing re-entry probes as well as assessing their performance in flight, with computational methods rather than direct ground testing facilities. The general framework of the testing methodology for high-speed re-entry is summarized in Fig. 9.12.

9.7 Conclusion

The reproduction of high-speed re-entry conditions in ground-based facilities is a real challenge. Because of the coupling phenomena that characterized this type of flows their accurate experimental simulation on scaled-down models is impossible in ground-based facilities. It appears as well that only little research has been conducted on dedicated strategies for ground testing of high-speed re-entry flows. Most of the time, testing is limited to qualification tests that reproduce the heat flux level without taking into account all the physical phenomena involved.

The two main limitations concern radiation and gas–surface interactions: both cannot be correctly reproduced at the same time in a single facility. Indeed, the time-scale achieved in impulse facilities is shorter than those relevant for gas–surface interactions, while the correct radiation phenomena are difficult to reproduce in plasma wind tunnels.

A solution is to develop models for radiation and gas–surface interaction in the relevant facilities, under controlled environments. Since the conditions in which those models can be validated in ground facilities are different from the one encountered in high-speed flows, they specifically require to be physic based in view of their extrapolation to flight conditions. In such a context, model validation and their incorporation in CFD codes are crucial for the development of aerospace applications. UQ methodologies are expected to play a major role for this approach as they represent the unique way to give solid basis to the validation process. In order to manifest all their benefit and correctly address the problem, UQ methods imperatively need to be articulated with the experimental procedure. To this end, this brief review exposes the rationale of experimental testing and how it is linked to the physics of aerospace flights. It would serve as a basis for the development of Uncertainty Quantification apply to the validation of high-speed flow modeling.

References

1. P. Barbante and O. Chazot. Flight extrapolation of plasma wind tunnel of stagnation region flowfield. *J. Thermophys. Heat Transfer*, 20(3):493–499, 2006.
2. P. L. Chambré and A. Acrivos. On chemical surface reaction in laminar boundary layer flows. *J. Appl. Phys.*, 27(11):1322–1328, 1956.
3. O. Chazot, F. Panerai, and V. Van Der Haegen. Aerothermochemistry testing for lifting reentry vehicles. In ESA, editor, *7th European Symposium on Aerothermodynamics*, 2011.
4. O. Chazot, R. Régnier, and A. Garcia Munõz. Simulation methodology in plasmatron facility and hypersonic wind tunnels. In *12th International Conference on Method of Aerophysical Research*, 2004.
5. G. de Crombrugghe, R. Morgan, and O. Chazot. Theoretical approach and experimental verification of diffusive transport under binary scaling conditions. *International Journal of Heat and Mass Transfer*, (97):675–682, 2016.
6. R. J.-A. Desideri, Glowinski, and J. Periaux. *Hypersonic Flows for Re-entry Problems*. Springer-Verlag, Berlin, 1991. Viviand, H., Similitude in Hypersonic Aerodynamics, pp.72-97.
7. D. Elligton. Binary scaling limits for hypersonic flight. *AIAA J.*, 5(9):1705–1706, 1967.

8. J.A. Fay and F.R Riddell. Theory of stagnation point heat transfer in dissociated air. *J. Aeron. Sci.*, 25(2):73–85, 1958.
9. N.C. Freeman. Non-equilibrium flow of an ideal dissociating gas. *J. Fluid Mech.*, 4(part 4), 1958.
10. R. Goulard. The coupling of radiation and convection in detached shock layers. *Journal of Quantitative Spectroscopy and Radiative Transfer*, 1.
11. W. D. Hayes and R. F Probstein. *Hypersonic Flow Theory*. Academic Press, 1959.
12. W. D. Hayes and R. F. Probstein. Viscous hypersonic similitude. *J. Aeron. Sci.*, 26(12):815–825, 1959.
13. H. Hornung. Experimental real gas hypersonics. *Aeronaut. J.*, 92:379–389, 1988.
14. Kolesnikov A.F. Condition of simulation of stagnation point heat transfer from high enthalpy flow. *Fluid Dynamics*, 28(1):131–137, 1993.
15. W. Kordulla, X. Bouis, and G. Eitelberg. Wind tunnels for space applications at dlr and onera. In *Proceeding AAAF, 12ᵗʰ European Aerospace Conference*, 1999.
16. F. K. Lu and D. E. Marren. *Advanced Hypersonic Test Facilities, Prog. in Astronaut. and Aeronaut.*, volume 198. AIAA, Reston, MA, 2002. Principles of Hypersonic Test Facility Development, pp. 17-25.
17. Parck C. Laboratory simulation of aerothermodynamic phenomena: a review. In AIAA, editor, *17th Aerospace Ground Testing Conference*, 1992.
18. Richard Morgan, Tim McIntyre, and al. Impulse facilities for the simulation of hypersonic radiating flows. In AIAA, editor, *38th Fluid Dynamics Conference and Exhibit*, 2008.
19. Rosner D.E. *Chemically frozen boundary layer with surface reaction*. Princeton University, Princeton N.J., 1958. March.
20. W. Saric, J. Muyalert, and C. Dujaric. Hypersonic experimental and computational capability. improvement and validation. Technical report, AGARD-AR-319, Vol. I, 1996.
21. E. Venkatapathy and al. Thermal protection system technologies for enabling future sample return missions. Technical report, NASA Ames Research Center, 2009.

Printed in the United States
by Baker & Taylor Publisher Services